T0213759

Lecture Notes in Physics

Volume 988

The Lecture Notes in Physics

The series Lecture Notes in Physics (LNP), founded in 1969, reports new developments in physics research and teaching-quickly and informally, but with a high quality and the explicit aim to summarize and communicate current knowledge in an accessible way. Books published in this series are conceived as bridging material between advanced graduate textbooks and the forefront of research and to serve three purposes:

- to be a compact and modern up-to-date source of reference on a well-defined topic;
- to serve as an accessible introduction to the field to postgraduate students and nonspecialist researchers from related areas;
- to be a source of advanced teaching material for specialized seminars, courses and schools.

Both monographs and multi-author volumes will be considered for publication. Edited volumes should however consist of a very limited number of contributions only. Proceedings will not be considered for LNP.

Volumes published in LNP are disseminated both in print and in electronic formats, the electronic archive being available at springerlink.com. The series content is indexed, abstracted and referenced by many abstracting and information services, bibliographic networks, subscription agencies, library networks, and consortia.

Proposals should be sent to a member of the Editorial Board, or directly to the responsible editor at Springer:

Dr Lisa Scalone
Springer Nature
Physics
Tiergartenstrasse 17
69121 Heidelberg, Germany
lisa.scalone@springernature.com

More information about this series at http://www.springer.com/series/5304

Ramona Wolf

Quantum Key Distribution

An Introduction with Exercises

 Springer

Ramona Wolf (iD)
Institute for Theoretical Physics
ETH Zurich
Zurich, Switzerland

ISSN 0075-8450   ISSN 1616-6361   (electronic)
Lecture Notes in Physics
ISBN 978-3-030-73990-4   ISBN 978-3-030-73991-1   (eBook)
https://doi.org/10.1007/978-3-030-73991-1

This Springer imprint is published by the registered company Springer Nature Switzerland AG.
The registered company address is: Gewerbestrasse 11, 6330 Cham, Switzerland

Preface

These lecture notes grew out of a course on quantum key distribution taught at Leibniz Universität Hannover during the summer term of 2020 as part of the physics master's programme. When preparing the course, I discovered that there are very few textbooks that introduce quantum cryptography and quantum key distribution from the ground up, especially with regard to recent developments both on the theoretical and on the experimental front. Therefore, although the course was originally intended as a master's course, I decided to prepare these notes for a wider audience, including PhD students and postdocs who are new to the field of quantum key distribution.

This book is intended to serve as an introduction to the fast-developing field of quantum key distribution. It requires only basic knowledge of quantum mechanics and linear algebra but no prior knowledge of quantum information theory. The mathematical tools needed are presented in the first part of the book, using a variety of worked-out examples and exercises. In particular, the topic of classical and quantum entropies is presented in detail since it is a crucial ingredient of security proofs of quantum key distribution protocols but usually not part of theoretical physics courses. The book furthermore presents a variety of quantum key distribution protocols and different techniques to prove their security. In addition, we discuss the advances and challenges that occur when abstract quantum key distribution protocols are turned into technological prototypes, demonstrating its value to state-of-the-art cryptography and communication.

Acknowledgements

I would like to thank Alexander Kranz, Renato Renner, René Schwonnek, and Christian Staufenbiel for helpful discussions and comments on the manuscript. Special thanks goes to Tobias J. Osborne for encouraging me to publish these notes.

Zurich, Switzerland
February 2021

Ramona Wolf

Contents

About the Author

Ramona Wolf is a postdoctoral scholar in the Quantum Information Theory group at ETH Zurich, Switzerland, working on quantum cryptography. She obtained her PhD in physics from Leibniz University Hanover, Germany. Ramona has gained a lot of teaching experience by tutoring several courses in theoretical physics and also teaching a full course on quantum key distribution. She has published a number of articles in peer-reviewed academic journals and supervised several bachelor's and master's projects.

Introduction

1

Abstract

Quantum key distribution addresses one of the society's most pressing concerns for secret and authenticated communication. This is achieved by exploiting the principles of quantum theory to establish a secret key between two distant parties whose security is guaranteed by the laws of physics. In the past years, there has been tremendous progress with regard to the design of novel protocols, the development of sophisticated techniques for security proofs, and even in-field implementations of technological prototypes.

1.1 Classical Cryptography

The concept of private key cryptography is as old as the desire to exchange secret messages with others. The first descriptions of secret codes can already be found in chronicles of the war between the Persians and the Greeks in the fifth century BC. Throughout history, especially in times of war, secret writing has often decided life and death, victory or defeat. Codes and ciphers have evolved through history, there have always been attempts to crack ciphers and attempts to come up with new and more secure ones. What all these codes have in common is that they can, in principle, be cracked. It might take an exponentially long time (which is the case for codes based on factoring prime numbers) but it is not impossible. A famous example of how cryptography has had an influence on the outcome of a war is how the British (most famously, Alan Turing) have cracked the Enigma code used by the Germans to communicate during World War II.

A very simple example of an encryption scheme is the *Caesar cipher*, which is named after Julius Caesar who used it in correspondence that was of military significance. The idea is simple: every letter of the alphabet is shifted by a certain number of letters. In its original form every letter was shifted three letters. The

© The Author(s), under exclusive license to Springer Nature Switzerland AG 2021
R. Wolf, *Quantum Key Distribution*, Lecture Notes in Physics 988,
https://doi.org/10.1007/978-3-030-73991-1_1

plain alphabet, which is used to write the original message, is then encrypted in the following way by the *cipher alphabet*:

Plain alphabet	a b c d e f g h i j k l m n o p q r s t u v w x y z
Cipher alphabet	D E F G H I J K L M N O P Q R S T U V W X Y Z A B C

Suppose, for example, that we want to send the message "meet me at the apple tree" to someone, but we do not want anyone else to be able to read the message. We can use the Caesar cipher to encrypt the message:

Plain text	m e e t m e a t t h e a p p l e t r e e
Cipher text	P H H W P H D W W K H D S S O H E U H H

At first glance, the ciphertext does not make any sense. But can we be completely sure that an adversary is unable to get the original message? Certainly not! For instance, if he knows about the Caesar cipher, he can just reverse the encryption process. But even if he does not know about this encryption scheme, there is a way to recover the original message, which works by exploiting the structure of the English language (or any language, for that matter). More precisely, one can analyse the frequency with which certain letters appear in a typical English sentence which is reflected in the frequency of certain letters in the ciphertext. In the English language, the letter "E" is the most frequent one. The message above is a very good example for this: The letter "E" appears seven times in a message of 20 letters, which corresponds to a frequency of 35%. This is even above the typical frequency of 12.7% for the letter "E" [6] in the English language, but this is due to the short nature of the message. Hence, an adversary can easily determine that the letter "H" in the ciphertext corresponds to the letter "E" in the plain text.

There are of course more complicated encryption schemes than the Caesar cipher: for instance, one could shuffle the letters of the alphabet randomly instead of just shifting them by a certain number. However, this scheme can be cracked in the same way as the Caesar cipher, simply by applying frequency analysis. Even more sophisticated schemes such as the Vigenère cipher, which could not be cracked for over three centuries, turned out to be not secure in the end: although the scheme to crack this cipher was more sophisticated than a simple frequency analysis, it was finally found. This shows an important point: people who use certain cryptographic schemes always have to be vigilant in case someone found a way to crack their scheme. We do not want to go into too much detail here,[1] but one point should have become clear: there is a great desire for a cryptographic scheme that is *unbreakable* (or, in more technical terms, *information-theoretically secure*), even if an adversary had all the computational power in the universe.

---

[1]If you are interested in the history of cryptography, great references are [4] and [6].

1.2 Provably Secure Cryptography

Surprisingly, it was not until the 1920s that someone came up with an encryption scheme that was indeed provably secure: the *Vernam cipher* [7], also called *one-time pad*. Before we explain how this encryption scheme works, let us have a closer look at what it is that we actually want and fix some technical terms. Consider the following setting: two parties, traditionally called Alice and Bob, want to exchange messages over a distance and they want to be sure that no one is able to read these messages. Usually, we call this hypothetical adversary Eve. Hence, Alice and Bob need a *key* to encrypt and decrypt messages in a way that Eve cannot gain any information about the messages.

First of all, we have to define what we actually mean when we use the word *key*. In principle, a key can be anything that encrypts a message: some set of rules that tells you how to replace the letters of the message for encryption (e.g., the Caesar cipher we have seen above or the so-called four-square which encrypts pairs of letters), an artificial set of symbols, or even some sort of mechanical device that encrypts your message (e.g., a scytale). Basically, we mean any kind of mechanism that can be used to hide the message we want to send. However, when we use the word key in these notes from now on, we always refer to a string of bits, i.e., a sequence of 0s and 1s. In the same fashion, whenever we say "message", we mean a message which is in the form of a bit string.

Suppose Alice and Bob both hold a key, i.e., they share identical bit strings. How can they use this key to safely encrypt, send, and decrypt messages? Here, the one-time pad encryption scheme comes into play. Consider the following scenario (see Fig. 1.1): Alice and Bob have successfully created a pair of keys S_A, S_B, each of length n, and want to use it to send a message $M = (M_1, M_2, \ldots, M_n)$ of length n from Alice and Bob. $M_1, M_2, \ldots, M_n$ represent the individual bits of the message given as a bit string. The protocol for this is as follows:

1. Alice encrypts the message M with the key S_A by doing binary addition, which means that each bit of the ciphertext is produced by binary adding a bit from the key to a bit from the message:

$$C_i = M_i \oplus (S_A)_i. \qquad (1.1)$$

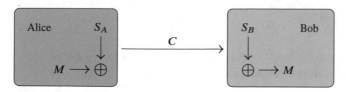

Fig. 1.1 One-time pad encryption. Alice encrypts a message $M \in \{0, 1\}^n$ of length n using the key $S_A \in \{0, 1\}^n$ by doing binary addition ($\oplus$). This results in the ciphertext $C \in \{0, 1\}^n$. Bob receives the ciphertext and decrypts it using his key $S_B \in \{0, 1\}^n$ performing binary addition. As a result, he gets the original message M (in the ideal case)

2. The ciphertext $C = (C_1, C_2, \ldots, C_n)$ is sent to Bob. This is done over a public channel, i.e., an adversary has full access to the ciphertext (but cannot change the message).
3. Bob decodes the message by doing binary addition of the ciphertext and his key S_B:

$$M_i = C_i \oplus (S_B)_i. \tag{1.2}$$

If the keys are equal, i.e., $S_A = S_B$, the message Bob holds after decoding is exactly the same as the one Alice sent[2] since

$$M_i = C_i \oplus (S_B)_i \tag{1.3}$$

$$= (M_i \oplus (S_A)_i) \oplus (S_B)_i \tag{1.4}$$

$$= M_i \oplus ((S_A)_i \oplus (S_B)_i) \tag{1.5}$$

$$= M_i \oplus ((S_A)_i \oplus (S_A)_i) \tag{1.6}$$

$$= M_i. \tag{1.7}$$

This encryption scheme is indeed information-theoretically secure, which means that the encrypted message (i.e., the ciphertext) provides no information about the original message (except the maximum possible length of the message), even though the adversary has complete access to the ciphertext. This was proved by Claude Shannon in 1949 [5], but the key has to fulfil certain properties:

1. It has to be truly random, which means that the individual key bits are not correlated in any way. Later, we will explicitly see why this is a crucial requirement.
2. It has to be at least as long as the original message, which follows directly from the encryption process described above.
3. It can never be reused in whole or partly. Otherwise, an adversary is able to get information about the messages that are sent, as can be seen in the following: suppose two messages M and N are encrypted using the same key S. If Eve has access to both resulting ciphertexts C_1 and C_2, she can simply compute

$$(C_1 \oplus C_2)_i = M_i \oplus S_i \oplus N_i \oplus S_i \tag{1.8}$$

$$= M_i \oplus N_i. \tag{1.9}$$

This does not directly reveal the messages that have been sent, but Eve certainly has *some* information about the messages now.

---

[2]Of course, we assume an ideal world without practical limitations as noise and losses here.

Fig. 1.2 Ideal key generator setting. An ideal device outputs two identical, random keys $S_A = S_B$ for Alice and Bob, while Eve cannot interact with it (i.e., nothing can be input our output to Eve's part of the system, which is denoted by the symbol $\perp$)

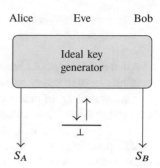

4. It has to be kept completely secret, i.e., the adversary has zero knowledge about the key.[3]

Although the one-time pad encryption scheme is provably information-theoretically secure, it obviously has some drawbacks with regard to practical usability. Generating a key that is truly random (i.e., building a device as depicted in Fig. 1.2) is a highly non-trivial task. Furthermore, Alice and Bob have to find a way to generate the exact same key in two distant places at once. If they had to meet to agree on the key, they could directly exchange the message they want to send.

Hence, what Alice and Bob want is a device that uses some sort of protocol (we do not care which one exactly) to generate keys for Alice and Bob. The scheme of such a device is depicted in Fig. 1.2. Since we cannot be a hundred percent certain that Eve does not have any access to this box, we also have to include her in the equation. What are the properties that we require of such an ideal device?

1. First of all, the key has to be *correct*, i.e., Alice and Bob should receive a matching set of keys in order to be able to communicate with each other:

$$S_A = S_B \equiv S, \tag{1.10}$$

where S can be thought of as a random variable (i.e., it has a probability distribution over all possible key values). The values that S can take are bit strings of length l (= the length of the key).
2. The key should be *random*, i.e., every possible bit string of length l appears with the same probability. Since there are 2^l such strings, we require that the probability that the key S takes a certain value s (i.e., a specific bit string) is

$$\Pr[S = s] = 2^{-l} \tag{1.11}$$

for any value s that the key S can take.

---

[3]This requirement is not always true. If Alice and Bob use the one-time pad encryption scheme and Eve learns about the key because she knew part of the message beforehand, this does not change the security of the scheme since the key is not reused.

3. Eve cannot interact with the device: she neither can input anything to it nor get any output.

If Alice and Bob would have access to such a device, they would be able to use the resulting keys to safely communicate, for example, via the one-time pad scheme.

Unfortunately, in practice we will never be able to construct such an ideal key generating device. For instance, we cannot completely prevent Eve from interacting with the box. There will probably be some information leaked to Eve, and in a practical setting she is also able to interact with the box, for instance, by simply cutting the fibres of the communication channels that Alice and Bob use. What can we do? We can weaken the requirements in a way that the resulting device is not perfect any more but its failure probability is small. The requirements for such a realistic device are then changed in the following way:

1. We do not require that the keys the device outputs are always equal, but that the probability for the protocol *not* to abort and Alice's key differs from Bob's key is very small:

$$\Pr[S_A \neq S_B] \leq \epsilon. \tag{1.12}$$

As usual in mathematics and physics, we take ϵ to be a very small number.
2. We now require that the generated key is *close* to a perfect key, which means that it is close to a uniformly random key. Furthermore, we require that the individual key bits do not have any correlation with each other,[4] i.e., knowing the first n key bits does not reveal any information about the $n + 1$-th key bit. Later, we will see why this latter condition is important to be able to safely use the key in any application. With regard to Eve's system, being close to a perfect key means that even though some information is leaked to Eve, the resulting key is independent of the state of Eve's system.
3. Although we cannot prevent Eve from attacking the device, for example, by cutting the fibre, we can detect such attacks. In these cases the device will abort the protocol and not output a key. In this way, we lose the property that the device always outputs a secure key, but we can be sure that *if* a key is produced, this key is secure.

Hence, the main goal of quantum key distribution is to build a device where, whenever a key is generated, we can be sure that it is secure according to the above (weakened) requirements, so that Alice and Bob never use a corrupted key. This is where quantum mechanics comes in: In the theory of quantum mechanics there are several phenomena that can be helpful to attack these problems: first of all, there is an inherent randomness to quantum states and measurements that one can exploit. Furthermore, quantum theory provides an amazingly practical result with respect to

---

[4]These statements will be made precise later.

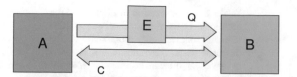

Fig. 1.3 Basic setting of the BB84 protocol. Two parties, Alice (A) and Bob (B), want to establish a secret key by using a quantum channel (Q) where an adversary, Eve (E), can tap into without restrictions and a classical channel (C) where Eve can only listen to the communication but cannot alter it

cryptography: the no-cloning theorem. Unlike classical physics, where states can be perfectly cloned arbitrarily often (at least, in principle), this is completely forbidden for quantum states due to the linearity of quantum theory.

To get an idea of how well the principles of quantum mechanics suit our needs in order to create a key that is usable for applications like one-time pad encryption, we study the first example of a quantum key distribution protocol: The BB84 protocol.

1.3   The BB84 Protocol

In 1984, Charles Bennett and Gilles Brassard developed the first quantum key distribution protocol [1, 2], called the *BB84 protocol*, which uses the encoding of classical bits into qubits, i.e., two-level quantum systems. In this example we will focus on the realization of qubits by using the polarization degree of photons.

The setting is the following: two authorized parties, called Alice and Bob, want to establish a secret key over a distance. To accomplish this task, they have access to two channels (see Fig. 1.3): The first one is a classical channel that Alice and Bob can use to send classical messages back and forth. It needs to be authenticated, which means that Alice and Bob identify themselves.[5] An adversary (Eve) can listen to the conversation but cannot change the messages that are sent. The second channel is a quantum channel that allows Alice and Bob to send quantum signals, which is completely insecure (i.e., the adversary can perform any operation on the information that is sent that is allowed by the rules of quantum mechanics).

The protocol can be divided into two parts: First, there is a *quantum transmission* part, where Alice and Bob prepare, send, and measure quantum states. The second part is the *classical post-processing* part which is purely classical. Alice and Bob only communicate over the classical channel to transform the bit strings they have obtained in the first phase into secure keys. Before we explain how the protocol works, we need to make a short detour to understand how we can use the polarization of photons for the encryption of information.

---

[5]Alice must be able to rely on the fact that it is really Bob she is communicating with (and vice versa). If the adversary just murders Bob in his lab and pretends to be him, there is no point in trying to generate a secret key.

Table 1.1 Polarization of photons. In the rectilinear basis, the photon is either horizontally or vertically polarized. In the diagonal basis, the polarization is +45° or −45°. A bit can be encoded into the polarization of a photon in either basis by assigning one polarization orientation to each of the respective values

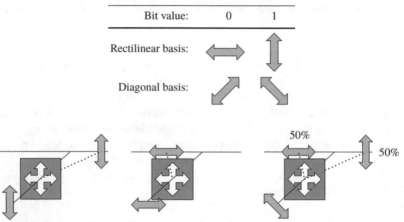

Fig. 1.4 Polarization filters. When a vertically (horizontally) polarized photon passes through a rectilinear polarization filter, it is deflected to the right (left), as depicted in the left and middle picture. When a photon that is polarized in the diagonal basis passes through a rectilinear polarization filter, it will be deflected to the left 50% of the time and to the right 50% of the time, see the picture on the right

1.3.1 Polarization of Photons

The polarization of light specifies the geometrical orientation of the oscillation of the electromagnetic field associated with its wave. It is perpendicular to the direction of the propagation of the light. We focus here on *linear polarization*, which means that the field only oscillates in one direction (in contrast to, e.g., circular polarization where the field rotates in a plane as the wave propagates).

We distinguish two kinds of bases of linear polarization states: the rectilinear basis, which includes horizontal and vertical orientations, and the diagonal basis, which includes orientations rotated by +45° and −45°. A classical bit (i.e., the values 0 and 1) can be encoded into the polarization of a photon as depicted in Table 1.1.

There exist filters to distinguish between horizontally and vertically polarized photons (and, analogously, between +45° and −45° rotated photons), as depicted in Fig. 1.4: when a vertically polarized photon passes through the filter, it is deflected to the right, while a horizontally polarized one is deflected to the left. Measuring a photon in one of these bases effectively means that the photon first passes the corresponding polarization filter and is detected by a photon detector afterwards (you need one detector for each orientation). But what happens when a diagonally polarized photon passes through the rectilinear filter? The photon then will be randomly deflected in one of the two directions. In this process, the polarization state of the photon is changed so that it is impossible to determine its orientation before passing the filter. Note that when measuring a diagonally polarized photon in the rectilinear basis you get a completely random result (and vice versa), which means that you do not gain any information. This means that the two bases are *mutually unbiased*.

1.3.2   Quantum Transmission

The quantum transmission phase includes all operations that are done on quantum states, which includes encoding and decoding of classical bits into quantum states and communication over a quantum channel.

1. Alice chooses a string of N random classical bits $X_1, \ldots, X_N$.
2. Alice chooses a random sequence of polarization bases, where she can either choose the *rectilinear basis* (R) $\oplus$ or the *diagonal basis* (D) $\times$ These two bases are *mutually unbiased*, i.e., a measurement in one of the bases reveals no information on a bit that is encoded in the other basis.
3. Alice encodes her bit string into a collection of photons with polarization according to the chosen bases, where the encoding works as shown in Table 1.1.
4. When Bob receives the photons he randomly (and independently of Alice!) decides for each photon whether to measure it in the rectilinear or the diagonal basis to obtain classical bits. After this step, Alice and Bob both hold a classical bit string, denoted $X = (X_1, \ldots, X_N)$ for Alice and $Y = (Y_1, \ldots, Y_N)$ for Bob. This is called the *raw key* pair.

1.3.3   Classical Post-Processing

The rest of the protocol is purely classical. Alice and Bob exchange a sequence of classical information to transform the bit strings they hold into a shared secret key.

5. Bob publicly announces the bases he has chosen to measure the photons Alice has sent. Alice compares Bob's bases to the ones she used and says which bases Bob has chosen correctly, i.e., in which cases their choices coincide. Alice and Bob discard all bits for which the encoding and measurement bases are not the same. This is called the *sifting step*.
6. The next step is the *parameter estimation* step, where Alice and Bob want to compute a guess for the *error rate* in the quantum channel, i.e., the fraction of

positions i where X_i and Y_i disagree. To achieve this, Bob reveals some bits of his key at random. In case of no eavesdropping, these bits should be the same as Alice's bits and she confirms them. If the error rate is too high, this indicates that there has been some eavesdropping and Alice and Bob abort the protocol. The bits that have been revealed during this step are discarded afterwards as their information is now public to an eavesdropper.

7. To compute the final key, Alice and Bob perform certain steps to correct errors in their keys and increase the secrecy of their key. These steps are called *error correction* (sometimes also referred to as information reconciliation), where they erase all errors in their bit strings, i.e., after this step Alice and Bob hold identical strings. The second step is *privacy amplification*, which is a procedure that minimizes Eve's knowledge of the key. These steps have not been discussed in the original proposal of the protocol and first appeared a few years later in [3]. We will go into more detail about these steps later.

After carrying out the steps of this protocol, Alice and Bob hold a pair of identical secure keys, which can then be used for tasks such as encrypting messages. In Table 1.2, an example of an implementation of the BB84 protocol is depicted.

Table 1.2 The BB84 protocol. The table shows an example for an implementation of the BB84 protocol using the polarization of photons to encode qubits, assuming that no eavesdropping takes place

Quantum transmission													
Alice's random bits	0	1	1	1	0	1	0	1	1	0	0	1	1
Random sending bases	D	R	D	D	R	D	R	R	D	D	D	R	D
Photons Alice sends	↗	↕	↘	↗	↔	↘	↔	↕	↘	↗	↗	↕	↘
Random receiving bases	D	R	R	D	R	D	D	D	R	D	R	R	D
Bits as received by Bob	0	1	0	1	0	1	1	1	0	0	0	1	1
Classical post-processing													
Bob reports bases of received bits	D	R	R	D	R	D	D	D	R	D	R	R	D
Alice says which bases are correct	OK	OK		OK	OK	OK				OK		OK	OK
Presumably shared information (without eavesdropper)	0	1		1	0	1				0		1	1
Bob reveals some key bits		1				1						1	
Alice confirms them		OK				OK						OK	
Outcome													
Remaining shared secret bits	0			1	0					0			1

1.3.4 Security of the BB84 Protocol

The security of the BB84 protocol relies on the fact that in quantum mechanics it is not possible to gain information about a quantum system without disturbing it. Hence, every interaction the eavesdropper has with a quantum state that is sent alters the state in some way. How can Alice and Bob recognize that an adversary is listening to their communication while performing the steps of the BB84 protocol? Let us have a look at what happens if Eve tries to gain information about the quantum state that Alice sends to Bob. Consider the first bit in the example of Table 1.2, but now Eve interacts with the quantum state:

Alice's bit:	0
Alice's basis:	D
State that is sent:	↗
Eve's measurement basis:	R
State after the measurement:	↔
Bob's measurement basis:	D
Bob's bit:	1

In this case, Bob and Alice have used the same bases, hence after the sifting step they believe that they are holding the same bit value. Eve's interaction, however, has changed the quantum state in a way that Bob's measurement has yielded a different bit than the one Alice encoded. In the parameter estimation step, where Bob and Alice compare parts of the actual bit strings they hold, they will find that these bits do not match and reveal that an eavesdropper has tried to get some information.

1.4 Structure of the Book

Throughout the whole book, exercises help to consolidate the material presented and to become more proficient in using the mathematical tools. The content of the book is organized as follows:

In Chap. 2 we give an introduction to the tools of quantum information theory that are needed to understand the remainder of the book. We discuss how quantum systems are described, how quantum measurements work and encounter quantum mechanical phenomena such as entanglement and the no-cloning theorem, which play an important role for the security of quantum key distribution (QKD).

In Chap. 3 we describe how information and the lack thereof can be quantified in terms of entropies. After presenting classical entropies and their properties we turn to their quantum counterparts, where we discuss both similarities and differences between the two. Furthermore, we study the uncertainty principle in terms of entropic quantities.

After having set the theoretical framework we describe QKD protocols in Chap. 4. Here, we discuss all stages of a typical QKD protocol: first, the quantum transmission phase, which can be done either with a prepare-and-measure scheme or with an entanglement-based one. We present several examples of protocols, such as the BB84 protocol, the SARG04 protocol, and the Ekert protocol. We then continue with the classical post-processing and the kind of techniques that are used to turn the raw bit strings into a secure key.

In Chap. 5 we focus on the security of QKD protocols. The first task here is to find a suitable definition of security for a QKD protocol and be clear about the kind of attacks an eavesdropper can perform. Afterwards, we present one of the first security proofs of the BB84 protocol by Shor and Preskill before we discuss modern techniques for security proofs such as the secret key rate and entropic uncertainty relations.

While QKD protocols suffer from the fact that practical devices never work exactly as specified in the theoretic protocol, device-independent QKD (DIQKD) circumvents this problem by not making any assumptions on the devices, which is presented in Chap. 6. We discuss device-independent concepts for quantum communication and how a QKD protocol that is based on these concepts can be proven to be secure. We also discuss loopholes that can be exploited to corrupt the security of a DIQKD scheme even though it makes no assumptions on the devices that are used.

In Chap. 7 we present recent developments in QKD with the focus on protocols that aim at overcoming practical challenges such as imperfect devices and losses in noisy channels as well as attacks tailored to exploit certain characteristics of detectors, for instance. We first present the concept of measurement device-independent QKD, which removes the need of trusting the measurement devices. We then introduce a version of QKD based on continuous variables (in contrast to discrete variables used in most protocols such as BB84). We end the chapter by giving an overview of the state-of-the-art experiments for all the different kind of protocols presented in this book.

References

1. Bennett, C.H., Brassard, G.: Quantum cryptography: public key distribution and coin tossing. Proc. IEEE Int. Conf. Comput. Syst. Sign. Process. **175**, 8 (1984)
2. Bennett, C.H., Brassard, G.: Quantum cryptography: public key distribution and coin tossing. Theoret. Comput. Sci. **56**, 7–11 (2014). https://doi.org/10.1016/j.tcs.2014.05.025
3. Bennett, C.H., Brassard, G., Robert, J.M.: Privacy amplification by public discussion. SIAM J. Comput. **17**(2), 210–229 (1988). https://doi.org/10.1137/0217014
4. Kahn, D.: The Codebreakers: The Story of Secret Writing. Scribner, New York (1996)
5. Shannon, C.E.: Communication theory of secrecy systems. Bell Syst. Tech. J. **28**(4), 656–715 (1949). https://doi.org/10.1002/j.1538-7305.1949.tb00928.x
6. Singh, S.: The Code Book. The Science of Secrecy from Ancient Egypt to Quantum Cryptography. Fourth Estate, London (1999)
7. Vernam, G.S.: Cipher printing telegraph systems: for secret wire and radio telegraphic communications. J. AIEE **45**(2), 109–115 (1926). https://doi.org/10.1109/jaiee.1926.6534724

Mathematical Tools

2

Abstract

To describe the systems and processes involved in a quantum key distribution protocol, we use the language of quantum mechanics. Therefore, we briefly recap the definitions and concepts that will be needed later, which includes the description of quantum systems, quantum phenomena such as entanglement, and measures of distance between quantum states.

For a more in-depth introduction to the mathematical methods of quantum information theory, we refer to a lecture series held by Reinhard Werner in 2017 [4] and the well-written and comprehensive textbook on quantum information theory by Mark Wilde [5]. Another great resource which also includes the field of quantum computation is the book by Nielsen and Chuang [3].

2.1 Basics in Linear Algebra

Before we discuss how to describe quantum mechanical systems, we recall some basic concepts of linear algebra such as Hilbert spaces and tensor products. This will form the mathematical framework for the description of quantum systems.

2.1.1 The Hilbert Space

When describing a quantum mechanical system, the starting point is usually the *Hilbert space* $\mathcal{H}$, which, according to the postulates of quantum mechanics, determines the state space of the system. We begin by recalling the basic definition and some important properties.

© The Author(s), under exclusive license to Springer Nature Switzerland AG 2021   13
R. Wolf, *Quantum Key Distribution*, Lecture Notes in Physics 988,
https://doi.org/10.1007/978-3-030-73991-1_2

Definition 2.1 A Hilbert space is a vector space over the complex numbers $\mathbb{C}^1$ equipped with a scalar product

$$(\cdot, \cdot) : \mathcal{H} \times \mathcal{H} \to \mathbb{C} \tag{2.1}$$

such that the following conditions are fulfilled:

1. $(\varphi, \psi) = \overline{(\psi, \varphi)}$ for all $\varphi, \psi \in \mathcal{H}$, where the bar denotes the complex conjugate.
2. $(\varphi, \lambda\psi + \eta) = \lambda\,(\varphi, \psi) + (\varphi, \eta)$ for all $\varphi, \psi, \eta \in \mathcal{H}$ and $\lambda \in \mathbb{C}.^2$
3. $(\varphi, \varphi) \leq 0$ for all $\varphi \in \mathcal{H}$, and $(\varphi, \varphi) = 0$ if and only if $\varphi = 0$.

In these notes we will use the *Dirac notation* or *bra-ket notation* to denote vectors in a Hilbert space, which means that a vector (a *ket*) in a Hilbert space $\mathcal{H}$ is denoted $|\varphi\rangle \in \mathcal{H}$. There exists a dual vector (a *bra*) $\langle\varphi| \in \mathcal{H}^*$ to every vector $|\varphi\rangle \in \mathcal{H}$, where $\mathcal{H}^*$ is the dual Hilbert space of $\mathcal{H}$. The scalar product of two vectors $|\varphi\rangle, |\psi\rangle \in \mathcal{H}$ is then denoted $\langle\varphi|\psi\rangle \equiv (\varphi, \psi)$.

Exercise 2.2 Show that $\langle\lambda\varphi|\psi\rangle = \overline{\lambda}\langle\varphi|\psi\rangle$.

The scalar product induces a norm on the Hilbert space which is defined as

$$\sqrt{\langle\varphi|\varphi\rangle} = \|\varphi\|. \tag{2.2}$$

From the properties of the scalar product it directly follows that this norm is non-negative and only zero whenever the vector itself is zero. The Hilbert space $\mathcal{H}$ is complete with respect to this norm.

Definition 2.3 An *orthonormal basis* of a Hilbert space $\mathcal{H}$ is a set of labelled vectors $\{|e_i\rangle\}_i$ such that

$$\langle e_i|e_j\rangle = \delta_{ij} \tag{2.3}$$

for all i, j and every vector $|\varphi\rangle \in \mathcal{H}$ can be written as a linear combination of the basis vectors in the following way:

$$|\varphi\rangle = \sum_i \langle e_i|\varphi\rangle|e_i\rangle. \tag{2.4}$$

---

[1] It is also possible to define a Hilbert space over the real numbers $\mathbb{R}$, but we do not care about this case in these lecture notes.

[2] Note that it is a convention to extract the complex conjugate out of the first factor and extract the factor linearly out of the second one. This convention is usually used by physicists, while the universal convention in mathematics is exactly the opposite.

Throughout these notes, we usually consider finite-dimensional Hilbert space, in which case the basis contains only finitely many elements.

2.1.2 Operators on Hilbert Spaces

Apart from the scalar product of two vectors, we can also calculate the *outer product* $|\varphi\rangle\langle\psi|$, which yields a linear operator on the Hilbert space $\mathcal{H}$ and acts on a vector $|\eta\rangle \in \mathcal{H}$ in the following way:

$$\big(|\varphi\rangle\langle\psi|\big)|\eta\rangle = |\varphi\rangle\langle\psi|\eta\rangle = \langle\psi|\eta\rangle|\varphi\rangle. \tag{2.5}$$

We can make use of the outer product to derive the completeness relation for orthonormal vectors. Suppose we have an orthonormal basis $\{|e_i\rangle\}_i$ of a Hilbert space $\mathcal{H}$. We can then write a vector $|\varphi\rangle \in \mathcal{H}$ as $|\varphi\rangle = \sum_i \varphi_i|e_i\rangle$ for some set of complex numbers φ_i. Note that $\varphi_i = \langle e_i|\varphi\rangle$, and hence

$$\left(\sum_i |e_i\rangle\langle e_i|\right)|\varphi\rangle = \sum_i |e_i\rangle\langle e_i|\varphi\rangle = \sum_i \varphi_i|e_i\rangle = |\varphi\rangle. \tag{2.6}$$

Since this is true for all vectors $|\varphi\rangle \in \mathcal{H}$, is follows that

$$\sum_i |e_i\rangle\langle e_i| = \mathbb{I}, \tag{2.7}$$

which is known as the completeness relation.

In general, a linear operator is defined as follows:

Definition 2.4 A linear operator between Hilbert spaces $\mathcal{H}$ and $\mathcal{H}'$ is a function $A : \mathcal{H} \to \mathcal{H}'$ which is linear in its inputs, i.e.,

$$A\left(\sum_i a_i|\varphi_i\rangle\right) = \sum_i a_i A\big(|\varphi_i\rangle\big). \tag{2.8}$$

An operator A on a Hilbert space $\mathcal{H}$, i.e., an operator $A : \mathcal{H} \to \mathcal{H}$, is said to be *bounded* if

$$\|A|\varphi\rangle\| \leq \|A\| \, \||\varphi\rangle\| \tag{2.9}$$

for all $|\varphi\rangle \in \mathcal{H}$ and the smallest possible constant $||A||$ that fulfils this equation is called the norm of the operator. We denote $\mathcal{B}(\mathcal{H})$ the set of all bounded linear operators acting on $\mathcal{H}$.[3]

Exercise 2.5 Prove the *Cauchy–Schwarz inequality*: $|\langle\varphi|\psi\rangle|^2 \leq \langle\varphi|\varphi\rangle\langle\psi|\psi\rangle$. *Hint: Make use of the completeness relation.*

Example 2.6 (Pauli Matrices) An important example of operators, which will play a big role in our discussion of quantum systems, is the *Pauli matrices*. They are operators on a two-dimensional Hilbert space and can therefore be written as 2×2 matrices:

$$X = \begin{pmatrix} 0 & 1 \\ 1 & 0 \end{pmatrix}, \quad Y = \begin{pmatrix} 0 & -i \\ i & 0 \end{pmatrix}, \quad Z = \begin{pmatrix} 1 & 0 \\ 0 & -1 \end{pmatrix}. \tag{2.10}$$

Together with the identity matrix $\mathbb{I}$ they form a basis for the 2×2 matrices.

An important function of operators is the *trace*. Let $A \in \mathcal{B}(\mathcal{H})$ and $\{|e_i\rangle\}$ be a basis of $\mathcal{H}$. The trace of A is defined as the sum of the diagonal matrix elements of A:

$$\mathrm{Tr}\,(A) = \sum_i \langle e_i|A|e_i\rangle. \tag{2.11}$$

It is independent of the chosen basis of the Hilbert space and satisfies some important properties:

1. It is linear: For $A, B \in \mathcal{B}(H)$, it holds that $\mathrm{Tr}\,(A + B) = \mathrm{Tr}\,(A) + \mathrm{Tr}\,(B)$ and $\mathrm{Tr}\,(\lambda A) = \lambda\mathrm{Tr}\,(A)$ for $\lambda \in \mathbb{C}$.
2. It fulfils the cyclic property: For $A, B, C \in \mathcal{B}(\mathcal{H})$, it holds that $\mathrm{Tr}\,(ABC) = \mathrm{Tr}\,(BCA) = \mathrm{Tr}\,(CAB)$.

Exercise 2.7 Show that each of the Pauli matrices has trace zero.

Adjoints
For an operator $A \in \mathcal{B}(\mathcal{H})$, the adjoint (or Hermitian conjugate) $A^\dagger$ of A is defined by

$$\left(A^\dagger|\varphi\rangle, |\psi\rangle\right) = \left(|\varphi\rangle, A|\psi\rangle\right). \tag{2.12}$$

---

[3]If $\mathcal{H}$ is finite-dimensional, every linear operator is automatically bounded and it can be expressed as multiplication by some fixed matrix.

An operator A is said to be *Hermitian* (or self-adjoint) if $A^\dagger = A$. It is called *unitary* if $AA^\dagger = A^\dagger A = \mathbb{I}$, where $\mathbb{I}$ is the identity matrix. It is called *normal* if $AA^\dagger = A^\dagger A$. Clearly, any operator that is Hermitian or unitary is also normal.

Exercise 2.8 Use the definition of the adjoint operator to show that for two operators $A, B \in \mathcal{B}(\mathcal{H})$, it holds that $(AB)^\dagger = B^\dagger A^\dagger$.

Exercise 2.9 For a vector $|\varphi\rangle \in \mathcal{H}$, we define $|\varphi\rangle^\dagger = \langle\varphi|$. Use this to show that $(A|\varphi\rangle)^\dagger = \langle\varphi|A^\dagger$.

Exercise 2.10 Show that the Pauli matrices are both Hermitian and unitary.

Projectors
An important class of Hermitian operators is the *projectors*:

Definition 2.11 Let $\mathcal{H}$ be a Hilbert space and $\mathcal{H}'$ a subspace of $\mathcal{H}$ with orthonormal basis $\{|e_i\rangle\}_i$. The projector of $\mathcal{H}$ onto the subspace $\mathcal{H}'$ is defined as

$$P_{\mathcal{H}'} = \sum_i |e_i\rangle\langle e_i|. \tag{2.13}$$

Exercise 2.12 Show that any projector squares to itself, i.e., $P^2 = P$ and that each eigenvalue of P is either 0 or 1.

Any self-adjoint operator A admits a *spectral decomposition*. Since the eigenvalues of a self-adjoint operator are real and eigenvectors that correspond to different eigenvalues are orthogonal, A can be written as

$$A = \sum_i \lambda_i P_i, \tag{2.14}$$

where the λ_i are the different eigenvalues and P_i is the orthogonal projection onto the subspace spanned by the eigenvectors that correspond to the eigenvalue λ_i.

Furthermore, every linear operator A admits a *singular value decomposition*. There exist unitary matrices U and V and a diagonal matrix D with non-negative, real entries such that

$$A = UDV. \tag{2.15}$$

The diagonal entries of D are called the *singular values* of A.

2.1.3   Composite Systems

The composite system of two Hilbert systems $\mathcal{H}_A$ and $\mathcal{H}_B$ is described by the tensor product Hilbert space $\mathcal{H}_A \otimes \mathcal{H}_B$. States in $\mathcal{H}_A \otimes \mathcal{H}_B$ are linear combinations of tensor products $|\varphi\rangle \otimes |\psi\rangle$ of elements $|\varphi\rangle \in \mathcal{H}_A$ and $|\psi\rangle \in \mathcal{H}_B$. In particular, if $\{|e_i\rangle\}_i$ is a basis for $\mathcal{H}_A$ and $\{|f_j\rangle\}_j$ is a basis for $\mathcal{H}_B$, the Hilbert space of the composite system, $\mathcal{H}_A \otimes \mathcal{H}_B$, has a basis $\{|e_i\rangle \otimes |f_j\rangle\}_{i,j}$. Therefore, the dimension of the composite Hilbert space is the product of the dimensions of the subsystem Hilbert spaces:

$$\dim\left(\mathcal{H}_A \otimes \mathcal{H}_B\right) = \dim\left(H_A\right)\dim\left(H_B\right). \tag{2.16}$$

The tensor product fulfils the following properties for an arbitrary scalar $\lambda \in \mathbb{C}$ and vectors $|\varphi\rangle, |\varphi_1\rangle, |\varphi_2\rangle \in \mathcal{H}_A$ and $|\psi\rangle, |\psi_1\rangle, |\psi_2\rangle \in \mathcal{H}_B$:

1. $\lambda\left(|\varphi\rangle \otimes |\psi\rangle\right) = \left(\lambda|\varphi\rangle\right) \otimes |\psi\rangle = |\varphi\rangle \otimes \left(\lambda|\psi\rangle\right)$,
2. $\left(|\varphi_1\rangle + |\varphi_2\rangle\right) \otimes |\psi\rangle = |\varphi_1\rangle \otimes |\psi\rangle + |\varphi_2\rangle \otimes |\psi\rangle$,
3. $|\varphi\rangle \otimes \left(|\psi_1\rangle + |\psi_2\rangle\right) = |\varphi\rangle \otimes |\psi_1\rangle + |\varphi\rangle \otimes |\psi_2\rangle$.

To get a vector representation of states in a composite system, we use the definition of the tensor product from linear algebra. Suppose we have the two two-dimensional vectors

$$\begin{pmatrix} a_1 \\ b_1 \end{pmatrix}, \quad \begin{pmatrix} a_2 \\ b_2 \end{pmatrix}. \tag{2.17}$$

The tensor product of these two vectors is given by

$$\begin{pmatrix} a_1 \\ b_1 \end{pmatrix} \otimes \begin{pmatrix} a_2 \\ b_2 \end{pmatrix} = \begin{pmatrix} a_1 \begin{pmatrix} a_2 \\ b_2 \end{pmatrix} \\ b_1 \begin{pmatrix} a_2 \\ b_2 \end{pmatrix} \end{pmatrix} = \begin{pmatrix} a_1 a_2 \\ a_1 b_2 \\ b_1 a_2 \\ b_1 b_2 \end{pmatrix}. \tag{2.18}$$

A short remark on the notation: Since tensor products of states occur very often, people have developed various shorthands for this. For a composite system of two states, the following notations are equivalent:

$$|e\rangle \otimes |f\rangle = |e\rangle|f\rangle = |ef\rangle. \tag{2.19}$$

These notations are used when there is only one party involved. When there are two or more parties, e.g., A and B, we can indicate that the first state is local to the Hilbert space $\mathcal{H}_A$ and the second is local to the Hilbert space $\mathcal{H}_B$ by adding the respective subscripts to the states:

$$|e\rangle_A \otimes |f\rangle_B = |e\rangle_A |f\rangle_B = |ef\rangle_{AB}. \tag{2.20}$$

We will use any of these notations throughout these notes.

2.2 Description of Quantum Systems

The usual setting of a quantum mechanical experiment can be divided into three parts as depicted in Fig. 2.1: A preparation step, where the initial state ρ of the system is prepared, an evolution part that is described by a quantum channel $\mathcal{E}$, and a measurement step, where some observable M is measured, which yields an outcome x. In the following, we describe each of these steps in detail and give all the definitions that are needed in the following chapters.

2.2.1 States in Hilbert Space

We begin the description of quantum mechanical systems with the preparation step, the left (red) box in Fig. 2.1. The state space of the system of interest is given by a Hilbert space $\mathcal{H}$. The state of the system is then a vector $|\psi\rangle \in \mathcal{H}$ in this Hilbert space. The simplest example of a quantum state is a two-state system, a *qubit*. The two states of the qubit are denoted by $|0\rangle$ and $|1\rangle$. These can, for example, correspond to the rectilinear basis in the BB84 protocol, i.e., vertically and horizontally polarized photons. Moreover, the system can be in an arbitrary superposition of the two states:

$$|\psi\rangle = \alpha|0\rangle + \beta|1\rangle, \tag{2.21}$$

where α and β are arbitrary complex numbers that fulfil $|\alpha|^2 + |\beta|^2 = 1$. The states $|0\rangle$ and $|1\rangle$ have a vector representation of the form

$$|0\rangle = \begin{pmatrix} 1 \\ 0 \end{pmatrix}, \qquad |1\rangle = \begin{pmatrix} 0 \\ 1 \end{pmatrix}. \tag{2.22}$$

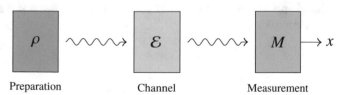

Fig. 2.1 General setting of a quantum mechanical experiment. This can be split into three parts: the preparation part, where the state of the system is prepared, the evolution that the system undergoes, and the measurement part, where some observable is measured

This makes it easy to see that they form an orthonormal basis for the space of states of a qubit. This basis is called the *computational basis*. In this basis, the state $|\psi\rangle$ is given by

$$|\psi\rangle = \begin{pmatrix} \alpha \\ \beta \end{pmatrix}. \tag{2.23}$$

Another important basis for qubit states in quantum information theory is the *Hadamard basis*, which is defined as follows:

$$|+\rangle = \frac{1}{\sqrt{2}} \begin{pmatrix} 1 \\ 1 \end{pmatrix}, \qquad |-\rangle = \frac{1}{\sqrt{2}} \begin{pmatrix} 1 \\ -1 \end{pmatrix}. \tag{2.24}$$

In the example of the BB84 protocol, the Hadamard basis corresponds to the diagonal basis. The relationship to the computational basis is the following:

$$|+\rangle = \frac{|0\rangle + |1\rangle}{\sqrt{2}}, \qquad |-\rangle = \frac{|0\rangle - |1\rangle}{\sqrt{2}}. \tag{2.25}$$

There is an operation that maps from the computational basis to the Hadamard basis, and vice versa, which is called the *Hadamard transformation H*:

$$H|0\rangle = \frac{|0\rangle + |1\rangle}{\sqrt{2}} = |+\rangle \tag{2.26}$$

$$H|1\rangle = \frac{|0\rangle - |1\rangle}{\sqrt{2}} = |-\rangle. \tag{2.27}$$

Its matrix representation is

$$H = \frac{1}{\sqrt{2}} \begin{pmatrix} 1 & 1 \\ 1 & -1 \end{pmatrix}. \tag{2.28}$$

It is easy to check that $H^2 = \mathbb{I}$, which implies that $H^{-1} = H$, hence H also maps the Hadamard basis to the computational basis.

A helpful visualization of the state of a qubit is given by the *Bloch sphere*, depicted in Fig. 2.2. Here, every quantum state $|\psi\rangle$ is represented by a point on the unit sphere. Suppose the amplitudes α and β in (2.21) have the following representation as complex numbers:

$$\alpha = r_0 e^{i\varphi_0} \tag{2.29}$$

$$\beta = r_1 e^{i\varphi_1}. \tag{2.30}$$

Fig. 2.2 Bloch sphere. Any qubit state $|\psi\rangle$ has a representation in terms of two angles $0 \leq \varphi \leq 2\pi$ and $0 \leq \theta \leq n$ given by $|\psi\rangle = \cos(\theta/2)|0\rangle + \sin(\theta/2)e^{i\varphi}|1\rangle$. The angles φ and θ determine a point on the unit sphere

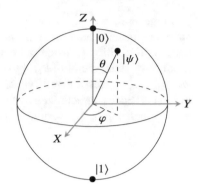

We can thus write the state $|\psi\rangle$ as

$$|\psi\rangle = \begin{pmatrix} r_0 e^{i\varphi_0} \\ r_1 e^{i\varphi_1} \end{pmatrix} = e^{i\varphi_0} \begin{pmatrix} r_0 \\ r_1 e^{i(\varphi_1 - \varphi_0)} \end{pmatrix}. \tag{2.31}$$

The states of two qubits are physically equivalent if they only differ by a global phase since this phase cannot be observed in a measurement (this will become clear when we discuss measurements of quantum systems). Hence, we can neglect the global phase $e^{i\phi_0}$ and the state $|\psi\rangle$ can be written

$$|\psi\rangle = r_0|0\rangle + r_1 e^{i(\varphi_1 - \varphi_0)}|1\rangle. \tag{2.32}$$

We now set $\varphi \equiv \varphi_1 - \varphi_0$, where $0 \leq \varphi \leq 2\pi$. Recall that since $|\psi\rangle$ is a quantum state, it has unit trace $\langle\psi|\psi\rangle$, which implies that $|r_0|^2 + |r_1|^2 = 1$. Therefore, the coefficients r_0 and r_1 can be parametrized by a single parameter θ:

$$r_0 = \cos(\theta/2) \tag{2.33}$$

$$r_1 = \sin(\theta/2), \tag{2.34}$$

with $0 \leq \theta \leq \pi$. As a result, we have rewritten the qubit state $|\psi\rangle$ in terms of the two angles φ and θ, which gives the Bloch sphere representation of the state:

$$|\psi\rangle = \cos(\theta/2)|0\rangle + \sin(\theta/2)e^{i\varphi}|1\rangle. \tag{2.35}$$

The angles φ and θ determine a point on the unit sphere (see Fig. 2.2). In Cartesian coordinates, this point is given by $(\cos\varphi\sin\theta, \sin\varphi\sin\theta, \cos\theta)$. This vector is also called the *Bloch vector*.

Exercise 2.13 Find the angles θ_+, φ_+ and θ_-, φ_- that determine the location on the Bloch sphere of the states $|+\rangle$ and $|-\rangle$.

In general, we may not have perfect knowledge of the prepared quantum state but instead have an ensemble of states, where each state occurs with a certain probability: Consider a quantum system which is described by a statistical mixture of state vectors $|\psi_1\rangle, |\psi_2\rangle, \ldots, |\psi_d\rangle \in \mathcal{H}$. The states have probabilities $p_1, p_2, \ldots, p_d$, respectively, which fulfil $p_i \geq 0$ and $\sum_i p_i = 1$. The state of the system is hence characterized by the *ensemble* $\{p_i, |\psi_i\rangle\}_{i=1,\ldots,d}$. The corresponding *density matrix* ρ of this system is defined as

$$\rho = \sum_{i=1}^{d} p_i |\psi_i\rangle\langle\psi_i|, \tag{2.36}$$

which can also be interpreted as the expectation of the states $|\psi_i\rangle$ with respect to the probability distribution given by the p_i. The state vectors $|\psi_i\rangle$ are normalized but they are not necessarily pairwise orthogonal. The formal definition of a density matrix is as follows:

Definition 2.14 A density matrix (or density operator) ρ is an operator on the Hilbert space $\mathcal{H}$ that satisfies the following properties:

1. It is normalized: $\mathrm{Tr}\,(\rho) = 1$.
2. It is Hermitian: $\rho^\dagger = \rho$.
3. It is positive semi-definite: $\langle\varphi|\rho|\varphi\rangle \geq 0$ for all $|\varphi\rangle$. We usually write $\rho \geq 0$ to indicate that an operator is positive semi-definite.

Note that although every ensemble has a unique density operator, the opposite does not necessarily hold. A given density operator can correspond to more than one ensemble. It is sometimes useful to work with states that are not normalized, i.e., $\mathrm{Tr}\,(\rho) \leq 1$. The set of subnormalized states on a Hilbert space $\mathcal{H}$ is denoted $S_\leq(\mathcal{H})$. Even though these states are not physical states, they can be renormalized to a physical state: If $\rho \in S_\leq(\mathcal{H})$, then $\rho/\mathrm{Tr}\,(\rho)$ is a valid density operator.

Example 2.15 A simple example for an ensemble is the following:

$$\left\{ \left\{ \frac{1}{5}, |0\rangle \right\}, \left\{ \frac{4}{5}, |1\rangle \right\} \right\}. \tag{2.37}$$

This can be interpreted as the system is in the state $|0\rangle$ with probability $\frac{1}{5}$ and in the state $|1\rangle$ with probability $\frac{4}{5}$. The resulting density matrix is

$$\rho = \frac{1}{5}|0\rangle\langle 0| + \frac{4}{5}|1\rangle\langle 1|. \tag{2.38}$$

Exercise 2.16 Given an ensemble $\{p_i, |\psi_i\rangle\}$, show that the resulting density operator ρ fulfils the properties 1.–3. stated in Definition 2.14.

Exercise 2.17 Consider the two ensembles $\left\{\{\frac{1}{2}, |0\rangle\}, \{\frac{1}{2}, |1\rangle\}\right\}$ and $\left\{\{\frac{1}{2}, |+\rangle\}, \{\frac{1}{2}, |-\rangle\}\right\}$. Show that they yield the same density matrix by applying the formula in (2.36).

Example 2.18 An important state, which will appear often throughout these notes, is the *maximally mixed state* π. It is a density operator corresponding to a uniform ensemble of orthogonal states $\left\{\frac{1}{d}, |x\rangle\right\}_{x=1,...,d}$, where d is the dimension of the Hilbert space. The maximally mixed state then corresponds to the density operator

$$\pi = \frac{1}{d} \sum_x |x\rangle\langle x| = \frac{\mathbb{I}}{d}. \tag{2.39}$$

For a single qubit, the maximally mixed state is a mixture of the two possible states $|0\rangle$ and $|1\rangle$ weighted with equal probability:

$$\pi_{\text{qubit}} = \frac{1}{2} \left(|0\rangle\langle 0| + |1\rangle\langle 1|\right) = \frac{\mathbb{I}}{2}. \tag{2.40}$$

If we know the state of the quantum system precisely, we say the system is in a *pure state*. In this case, the sum in (2.36) has only one term and the density matrix ρ is a one-dimensional projector. For instance, suppose the system is in the state $|\psi_3\rangle$. We then have $p_3 = 1$ and $p_i = 0 \; \forall \; i \neq 3$. Hence, the resulting density matrix is

$$\rho = |\psi_3\rangle\langle\psi_3|. \tag{2.41}$$

If there is more than one term in the sum, the system is said to be in a *mixed state*. Equivalently, we can use the *purity* to distinguish pure and mixed states.

Definition 2.19 The purity $\mathcal{P}(\rho)$ of a density matrix ρ is given by

$$\mathcal{P}(\rho) = \text{Tr}\left(\rho^\dagger \rho\right) = \text{Tr}\left(\rho^2\right). \tag{2.42}$$

A density matrix ρ is pure if $\mathcal{P}(\rho) = 1$ and it is mixed if $\mathcal{P}(\rho) < 1$. Subsequently, we will use the word *state* to refer to a general density matrix and the word *pure state* to refer to both the state vector $|\psi\rangle$ as well as the corresponding density matrix $\rho = |\psi\rangle\langle\psi|$.

By the spectral theorem it is possible to choose vectors $\{|\varphi_i\rangle\}_{i=1,...,d}$ that are pairwise orthogonal eigenvectors of ρ with corresponding eigenvalues λ_i such that

$$\rho = \sum_{i=1}^{d} \lambda_i |\varphi_i\rangle\langle\varphi_i|. \tag{2.43}$$

Thus, given a density operator ρ we can define its corresponding canonical ensemble $\{\lambda_i, |\varphi_i\rangle\}$. Note that this decomposition is not unique: If two eigenvalues are the same, i.e., $\lambda_i = \lambda_j$ for some $i \neq j$, then the choice of eigenvectors corresponding to these eigenvalues is not unique.

Exercise 2.20 Let $\rho = \sum_x \lambda_i |\varphi_i\rangle\langle\varphi_i|$ be the spectral decomposition of a density matrix $\rho \in \mathcal{H}$. Show that $\mathcal{P}(\rho) = 1$ if and only if ρ is a pure state, i.e., the spectral decomposition of ρ consists of only one term, and $\mathcal{P}(\rho) < 1$ otherwise.

Bloch Sphere for Mixed States

The Bloch sphere representation of pure single qubit states can be generalized to mixed qubit states. To recall the definition of the *Pauli matrices*:

$$X = \begin{pmatrix} 0 & 1 \\ 1 & 0 \end{pmatrix}, \quad Y = \begin{pmatrix} 0 & -i \\ i & 0 \end{pmatrix}, \quad Z = \begin{pmatrix} 1 & 0 \\ 0 & -1 \end{pmatrix}. \tag{2.44}$$

Using these, we can write any density operator as

$$\rho = \frac{1}{2}\left(\mathbb{I} + r_x X + r_y Y + r_z Z\right). \tag{2.45}$$

The vector $\mathbf{r} = (r_x, r_y, r_z)$ determines the point that corresponds to ρ. These points now do not only live on the unit sphere, but also in the interior of the sphere, called the *Bloch ball*.

Exercise 2.21

1. Show that the matrix representation of the state in (2.45) is

$$\rho = \frac{1}{2}\begin{pmatrix} 1 + r_z & r_x - ir_y \\ r_x + ir_y & 1 - r_z \end{pmatrix}. \tag{2.46}$$

2. Show that this state fulfils the properties of a density matrix stated in Definition 2.14.
3. Show that a state ρ is pure if and only if $||\mathbf{r}|| = 1$.
4. Calculate the Bloch vector for the maximally mixed state.

2.2.2 Quantum Channels

The dynamics of a quantum mechanical system is described by a quantum channel, which is an operator that maps quantum states in a Hilbert space $\mathcal{H}_A$ to quantum states in a Hilbert space $\mathcal{H}_B$. In Fig. 2.1, this is depicted as the middle (green) box and the notation we usually use to denote a quantum channel is $\mathcal{E}$. In the following, we will follow an axiomatic approach to determine the properties of a quantum channel. The properties we request are the following:

1. **Linearity.** The first requirement we impose on a quantum channel is that it should be a linear map:

Definition 2.22 Let $\mathcal{H}_A$ and $\mathcal{H}_B$ be Hilbert spaces. A map $\mathcal{E} : \mathcal{B}(\mathcal{H}_A) \to \mathcal{B}(\mathcal{H}_B)$ is said to be linear if

$$\mathcal{E}(\alpha \rho_A + \beta \sigma_A) = \alpha \mathcal{E}(\rho_A) + \beta \mathcal{E}(\sigma_A) \tag{2.47}$$

for density operators $\rho_A, \sigma_A \in \mathcal{B}(\mathcal{H}_A)$ and $\alpha, \beta \in \mathbb{C}$.

2. **Complete positivity.** First, it is clear that a quantum channel should preserve the class of positive semi-definite operators in order to take density matrices to density matrices. This means that it should be a positive map, which is defined as follows:

Definition 2.23 A linear map $\mathcal{E} : \mathcal{B}(\mathcal{H}_A) \to \mathcal{B}(\mathcal{H}_B)$ is positive if $\mathcal{E}(\rho_A)$ is positive semi-definite for all positive semi-definite $\rho_A \in \mathcal{B}(\mathcal{H}_A)$.

However, positivity is not sufficient for our purposes. Consider a variation of Fig. 2.1, where an additional "bystander", modelled by the Hilbert space $\mathcal{H}_{by}$, is involved in the experiment which is not affected by the channel (as depicted in Fig. 2.3). This situation can arise, for example, within an experiment where only a part of the prepared state undergoes a certain evolution. Although this differs from the original scenario, this way of operating on a subsystem should still be well-defined. In particular, we still require that the initial state $\rho \in \mathcal{H}_A \otimes \mathcal{H}_{by}$ is mapped to a density operator by the map $\mathcal{E} \otimes \mathrm{id}_{by}$, where id_{by} denotes the identity

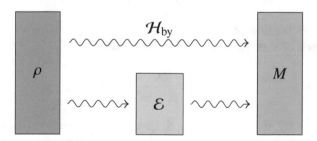

Fig. 2.3 A quantum mechanical system with a bystander. Here, we include an additional party, modelled by the Hilbert space $\mathcal{H}_{by}$, which is left unaffected by the quantum channel $\mathcal{E}$. The resulting quantum evolution $\mathcal{E} \otimes \mathrm{id}_{\mathcal{H}_{by}}$ still has to be positive

operator on the space $\mathcal{H}_{by}$. Hence, we require not only positivity, but *complete positivity*:

Definition 2.24 A linear map $\mathcal{E} : \mathcal{B}(\mathcal{H}_A) \to \mathcal{B}(\mathcal{H}_B)$ is completely positive if the map $\mathcal{E} \otimes \mathrm{id}_n : \mathcal{B}(\mathcal{H}_A) \otimes \mathcal{B}(\mathbb{C}^n) \to \mathcal{B}(\mathcal{H}_B) \otimes \mathcal{B}(\mathbb{C}^n)$ is positive for all $n \in \mathbb{N}$, where id_n denotes the identity map in $\mathbb{C}^n$.

3. **Trace Preserving.** The last requirement that we impose for quantum evolutions is trace preservation, which is again necessary to ensure that a quantum channel maps density operators to density operators. This means that the trace of the state does not change under the channel:

$$\mathrm{Tr}\,(\rho_A) = \mathrm{Tr}\,(\mathcal{E}(\rho_A)) \tag{2.48}$$

for all $\rho_A \in \mathcal{B}(\mathcal{H}_A)$.

Summing up the above discussion, we can define a quantum channel in the following way:

Definition 2.25 A quantum channel $\mathcal{E} : \mathcal{H}_A \to \mathcal{H}_B$ is a linear, completely positive, trace-preserving map.

Why is complete positivity an important requirement for quantum operations? An example of an operation on density matrices which is positive, but not completely positive, is the transpose operation on a single qubit. In the computational basis, this map is given by

$$\mathcal{T} : \rho \to \rho^T \tag{2.49}$$

$$\begin{pmatrix} a & b \\ c & d \end{pmatrix} \to \begin{pmatrix} a & c \\ b & d \end{pmatrix} \tag{2.50}$$

and it preserves the positivity of the single qubit density matrix. What happens if the qubit is part of a larger system, for example, in the state

$$|\Phi^+\rangle = \frac{|00\rangle + |11\rangle}{\sqrt{2}}? \tag{2.51}$$

In terms of density matrices, applying the transpose operation on the first qubit of the above state has the following effect on the density operator $\rho_{\Phi^+} = |\Phi^+\rangle\langle\Phi^+|$:

$$\frac{1}{2}\begin{pmatrix} 1 & 0 & 0 & 1 \\ 0 & 0 & 0 & 0 \\ 0 & 0 & 0 & 0 \\ 1 & 0 & 0 & 1 \end{pmatrix} \xrightarrow{\mathcal{T} \otimes \mathrm{id}} \frac{1}{2}\begin{pmatrix} 1 & 0 & 0 & 0 \\ 0 & 0 & 1 & 0 \\ 0 & 1 & 0 & 0 \\ 0 & 0 & 0 & 1 \end{pmatrix}. \tag{2.52}$$

The eigenvalues of the matrix after applying the map are $\frac{1}{2}, \frac{1}{2}, \frac{1}{2}$, and $-\frac{1}{2}$, hence it is not a positive matrix anymore and therefore not a valid density operator. This example shows that an operation which is positive, but not completely positive, can transform valid density operators into matrices that do not fulfil the properties of a density operator.

A quantum channel can equivalently be described by its *Kraus decomposition* due to the Choi-Kraus theorem:

Theorem 2.26 *A map* $\mathcal{E} : \mathcal{H}_A \rightarrow \mathcal{H}_B$ *is linear, completely positive, and trace-preserving if and only if it has a Kraus decomposition as follows:*

$$\mathcal{E}(\rho_A) = \sum_{j=1}^{d} K_j \rho_A K_j^\dagger, \tag{2.53}$$

where $\rho_A \in \mathcal{B}(\mathcal{H}_A)$, $K_j : \mathcal{H}_A \rightarrow \mathcal{H}_B$ *for all* $j \in \{1, \dots, d\}$ *and*

$$\sum_{j=1}^{d} K_j^\dagger K_j = \mathbb{I}_A, \tag{2.54}$$

and d need not by any larger than $\dim(\mathcal{H}_A) \dim(\mathcal{H}_B)$.

The full proof of this theorem can be found in [5, Theorem 4.4.1].

In a closed quantum system, the dynamics are described by a unitary operator $U : \mathcal{H} \rightarrow \mathcal{H}$, i.e., if ρ is the initial state of the system, then the state ρ' after the evolution is given by

$$\rho' = U\rho U^\dagger \equiv \mathcal{U}(\rho). \tag{2.55}$$

This kind of quantum evolution has only one Kraus operator, which is the unitary operator U. It is easy to reverse such a channel: the adjoint map $\mathcal{U}^\dagger$ is a unitary channel, and it can easily be shown that it is the inverse of $\mathcal{U}$:

$$(\mathcal{U}^\dagger \circ \mathcal{U})(\rho) = U^\dagger U \rho U^\dagger U = \rho. \tag{2.56}$$

A more general form is the *isometric channel* $\mathcal{V}(\rho) = V \rho V^\dagger$, where $V : \mathcal{H}_A \rightarrow \mathcal{H}_B$ is an isometry and $\dim(\mathcal{H}_A) \leq \dim(\mathcal{H}_B)$. Since it is an isometry, it satisfies $V^\dagger V = \mathbb{I}_{\mathcal{H}_A}$ but not, in general, $V V^\dagger = \mathbb{I}_{\mathcal{H}_B}$. It generalizes the unitary channel in the sense that it maps between spaces of different dimensions. This channel can also be reversed.

Exercise 2.27 In case of the isometric channel $\mathcal{V} : \mathcal{H}_A \rightarrow \mathcal{H}_B$ given by $\mathcal{V}(\rho_A) = V\rho_A V^\dagger$ with V being an isometry, the reverse channel is not given by the adjoint map $\mathcal{V}^\dagger$. However, this channel can still be reversed.

1. Show that $\mathcal{V}^\dagger$ is not a quantum channel in the sense of Definition 2.25.
2. Show that the map $\mathcal{R}$ defined as

$$\mathcal{R}(\rho_B) = \mathcal{V}^\dagger(\rho_B) + \mathrm{Tr}\left(\left(\mathrm{id}_{\mathcal{H}_B} - VV^\dagger\right)\rho_B\right)\sigma_A \tag{2.57}$$

with $\rho_B \in \mathcal{H}_B$ and $\sigma_A \in \mathcal{H}_A$ is completely positive and trace-preserving.
3. Show that the map $\mathcal{R}$ is the inverse of the isometric channel $\mathcal{V}$, i.e., that $(\mathcal{R} \circ \mathcal{V})(\rho_A) = \rho_A$.

Example 2.28 An example of a quantum channel that describes an irreversible evolution is the *amplitude damping channel*. To give it a physical interpretation, consider a two-level atom whose states are described in the computational basis: The ground state is described by $|0\rangle$ and the excited state is described by $|1\rangle$. The amplitude damping channel, depicted in Fig. 2.4, then describes the spontaneous decay of the atom from its excited state to the ground state, which occurs with a probability γ, where $0 \leq \gamma \leq 1$. This is even applicable if the atom is in superposition of the ground state and the excited state.

How can we find the Kraus operators for this channel? The operator that captures the decaying behaviour is the following:

$$K_1 = \sqrt{\gamma}|0\rangle\langle 1|, \tag{2.58}$$

since it decays the excited state to the ground state:

$$K_1|1\rangle\langle 1|K_1^\dagger = \gamma|0\rangle\langle 1|1\rangle\langle 1|1\rangle\langle 0| = \gamma|0\rangle\langle 0|. \tag{2.59}$$

This operator is not sufficient for the Kraus decomposition since it does not fulfil (2.54):

$$K_1^\dagger K_1 = \gamma|1\rangle\langle 1| \neq \mathbb{I}. \tag{2.60}$$

Fig. 2.4 The amplitude damping channel. With a probability γ, the atom decays from its excited state $|1\rangle\langle 1|$ to its ground state $|0\rangle\langle 0|$

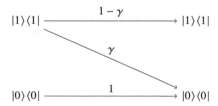

However, we can find the second Kraus operator by simply choosing an operator that, together with K_1, fulfils (2.54):

$$K_2^\dagger K_2 = \mathbb{I} - K_1^\dagger K_1 \tag{2.61}$$

$$= |0\rangle\langle 0| + |1\rangle\langle 1| - \gamma |1\rangle\langle 1| \tag{2.62}$$

$$= |0\rangle\langle 0| + (1 - \gamma)|1\rangle\langle 1|. \tag{2.63}$$

This condition is fulfilled by the operator

$$K_2 = |0\rangle\langle 0| + \sqrt{1 - \gamma}|1\rangle\langle 1|. \tag{2.64}$$

Hence, the amplitude damping channel can be represented by two Kraus operators.

Exercise 2.29 Show that applying the amplitude damping channel to a general quantum state ρ is equivalent to the following transformation of the Bloch vector:

$$\left(r_x, r_y, r_z\right) \to \left(r_x\sqrt{1 - \gamma}, r_y\sqrt{1 - \gamma}, \gamma + r_z(1 - \gamma)\right). \tag{2.65}$$

Exercise 2.30 The depolarizing channel is a qubit channel that describes a scenario where we completely lose the input qubit with some probability p. In this case, the input state is replaced by the maximally mixed state. The channel is defined as

$$\mathcal{E}(\rho) = (1 - p)\rho + p\pi, \tag{2.66}$$

where $\pi = \mathbb{I}/2$ is the maximally mixed state of a qubit system.

1. Show that for an arbitrary qubit state ρ, the identity $\mathbb{I}$ can be written as

$$\mathbb{I} = \frac{1}{2}\left(\rho + X\rho X + Y\rho Y + Z\rho Z\right), \tag{2.67}$$

where the X, Y, Z are the Pauli matrices.
2. Use the formula in (2.67) to show that the Kraus operators of the depolarizing channel are given by

$$K_1 = \sqrt{1 - \frac{3p}{4}}\ \mathbb{I}, \qquad K_2 = \frac{\sqrt{p}}{2}\ X,$$

$$K_3 = \frac{\sqrt{p}}{2}\ Y, \qquad K_4 = \frac{\sqrt{p}}{2}\ Z.$$

2.2.3 Measurements

At the end of an experiment, one usually makes an observation or a measurement
of the system to gain some sort of information about it, which is represented by the
right (blue) box in Fig. 2.1.

Definition 2.31 In general, a measurement is described by a collection $\{M_x\}$ of
operators, where $M_x : \mathcal{B}(\mathcal{H}) \rightarrow \mathcal{B}(\mathcal{H})$. The index x refers to the possible measurement outcomes, which are given by a finite outcome set X. The measurement
operators satisfy the *completeness relation*

$$\sum_{x \in X} M_x^\dagger M_x = \mathbb{I}. \tag{2.68}$$

If the system is in a pure state $|\psi\rangle$ when the measurement takes place, the
probability to obtain an outcome $x \in X$ is given by

$$P_\psi(x) = \langle \psi | M_x^\dagger M_x | \psi \rangle, \tag{2.69}$$

and the state $|\psi'\rangle$ of the system after the measurement is

$$|\psi'\rangle = \frac{M_x |\psi\rangle}{\sqrt{\langle \psi | M_x^\dagger M_x | \psi \rangle}}. \tag{2.70}$$

A simple yet important example of a measurement is the measurement of a single
qubit in the computational basis. This measurement has two possible outcomes, 0
and 1, and the corresponding measurement operators are given by

$$M_0 = |0\rangle\langle 0|, \qquad M_1 = |1\rangle\langle 1|. \tag{2.71}$$

Exercise 2.32 Show that the measurement operators defined in (2.71) fulfil the
completeness relation (2.68).

Exercise 2.33 Show that the measurement operators defined in (2.71) fulfil the
following relations:

$$M_0^\dagger M_0 = M_0^2 = M_0 \tag{2.72}$$

$$M_1^\dagger M_1 = M_1^2 = M_1. \tag{2.73}$$

Suppose we have a single qubit in the state $|\psi\rangle = \alpha|0\rangle + \beta|1\rangle$. According to (2.69), the probability of obtaining the outcome 0 is given by

$$P_\psi(0) = \langle\psi|M_0^\dagger M_0|\psi\rangle \tag{2.74}$$

$$= \langle\psi|M_0|\psi\rangle \tag{2.75}$$

$$= (\alpha^*\langle 0| + \beta^*\langle 1|)|0\rangle\langle 0|(\alpha|0\rangle + \beta|1\rangle) \tag{2.76}$$

$$= |\alpha|^2. \tag{2.77}$$

In the same way, we find that the probability of obtaining the outcome 1 is $P_\psi(1) = |\beta|^2$. For instance, if the system is in state $|\psi\rangle = \frac{|0\rangle+|1\rangle}{\sqrt{2}}$, we have that $P_\psi(0) = P_\psi(1) = \frac{1}{2}$, i.e., both outcomes occur with probability $\frac{1}{2}$. After the measurement, the system is in one of two states, depending on the measurement outcome:

$$|\psi_0'\rangle = \frac{M_0|\psi\rangle}{\sqrt{\langle\psi|M_0^\dagger M_0|\psi\rangle}} = \frac{\alpha}{|\alpha|}|0\rangle, \tag{2.78}$$

$$|\psi_1'\rangle = \frac{M_1|\psi\rangle}{\sqrt{\langle\psi|M_1^\dagger M_1|\psi\rangle}} = \frac{\beta}{|\beta|}|1\rangle. \tag{2.79}$$

We can ignore factors of the form $\alpha/|\alpha|$, which have modulus one, since they yield the same measurement statistics. More generally, we say that two states $|\psi\rangle$ and $e^{i\theta}|\psi\rangle$, where θ is a real number, are equal up to the global phase factor $e^{i\theta}$. This factor cannot be observed in a measurement: Given a measurement operator M_x, the respective probabilities of obtaining an outcome x for the two states are $\langle\psi|M_x^\dagger M_x|\psi\rangle$ and $\langle\psi|e^{-i\theta}M_x^\dagger M_x e^{i\theta}|\psi\rangle = \langle\psi|M_x^\dagger M_x|\psi\rangle$. Hence, there is no measurement that can distinguish between these two states, which is the reason why we regard global phase factors as irrelevant to the observed properties of the system.

The description of measurements can easily be extended to the density matrix formalism. Suppose the system is described by the ensemble $\{p_i, |\psi_i\rangle\}$, hence the corresponding density operator is given by

$$\rho = \sum_i p_i|\psi_i\rangle\langle\psi_i|. \tag{2.80}$$

We now want to perform a measurement given by the collection of operators $\{M_x\}$. If the system is in the state $|\psi_i\rangle$, we can rewrite the formula for the probability of obtaining the outcome x given in (2.69) as

$$P_{\psi_i}(x) = \langle\psi_i|M_x^\dagger M_x|\psi_i\rangle = \mathrm{Tr}\left(M_x^\dagger M_x|\psi_i\rangle\langle\psi_i|\right), \tag{2.81}$$

where we have used the cyclic property of the trace in the last step. The probability of obtaining the measurement outcome x, considering the complete ensemble is then (according to the law of total probability, see Appendix A) given by

$$P_\rho(x) = \sum_i p_i P_{\psi_i}(x) \tag{2.82}$$

$$= \sum_i p_i \text{Tr}\left(M_x^\dagger M_x |\psi_i\rangle\langle\psi_i|\right) \tag{2.83}$$

$$= \text{Tr}\left(M_x^\dagger M_x \rho\right). \tag{2.84}$$

The state of the system after obtaining the measurement outcome x is given by

$$\rho_x' = \frac{M_x \rho M_x^\dagger}{\text{Tr}\left(M_x^\dagger M_x \rho\right)}. \tag{2.85}$$

A special case of a measurement is the *projective measurement* defined as follows:

Definition 2.34 A projective measurement is a Hermitian operator $M : \mathcal{B}(\mathcal{H}) \to \mathcal{B}(\mathcal{H})$ with spectral decomposition

$$M = \sum_x x P_x, \tag{2.86}$$

where the x are the eigenvalues of M and the P_x form a set of orthogonal projectors satisfying $\sum_x P_x = \mathbb{I}$ and $P_x P_{x'} = \delta_{xx'} P_x$.

Projective measurements can be understood as a special case of the previously described measurements in the following way: Suppose we have a set of measurement operators $\{M_x\}$ which, in addition to the completeness relation, also satisfy the condition that they are orthogonal projectors, which means that they are Hermitian and fulfil $M_x M_{x'} = \delta_{xx'} M_x$. In this case, the M_x define a projective measurement M via $M = \sum_x x M_x$.

Instead of giving the observable M, one can also give the complete set of projective measurements $\{P_x\}$ to describe the measurement. The probability of obtaining an outcome x then simplifies to

$$P_\psi(x) = \langle\psi| P_x |\psi\rangle \tag{2.87}$$

and the state after the measurement is

$$|\psi'\rangle = \frac{P_x |\psi\rangle}{\sqrt{\langle\psi| P_x |\psi\rangle}}. \tag{2.88}$$

Exercise 2.35 Suppose the observable we want to measure is the Pauli-Z operator as defined in (2.44), which has eigenvalues $+1$ and -1 and eigenvectors $|0\rangle$ and $|1\rangle$.

1. What is the set of projection operators that corresponds to the Z measurement?
2. Suppose the initial state of the qubit is $|\psi\rangle = \frac{|0\rangle - |1\rangle}{\sqrt{2}}$. What is the probability of obtaining the outcomes $+1$ and -1, respectively?
3. Suppose the state of the system is given by the density matrix

$$\rho = \begin{pmatrix} 1/2 & 1/6 \\ 1/6 & 1/2 \end{pmatrix}. \tag{2.89}$$

   What is the probability of obtaining the outcome $+1$ when measuring the Z operator?

Exercise 2.36 Consider the following scenario, which typically appears in the BB84 protocol: Suppose Alice prepares a 0-bit in the computational basis, i.e., she sends the state $\rho = |0\rangle\langle 0|$ to Bob.

1. Suppose Bob chooses to measure the state also in the computational basis. What is the probability of getting the outcome 0 or 1, respectively?
2. Instead of measuring in the computational basis, Bob can also choose to measure the state in the Hadamard basis $\{|+\rangle, |-\rangle\}$, where the measurement operators are given by

$$E_+ = |+\rangle\langle +|, \qquad E_- = |-\rangle\langle -|. \tag{2.90}$$

   Show that the probabilities of getting the outcome $+$ and getting the outcome $-$ are the same when measuring the state $\rho = |0\rangle\langle 0|$, and hence Bob does not gain any information about the bit that Alice sent when measuring in the Hadamard basis.

An important problem in quantum information theory is distinguishing quantum states via measurements. If we have a set of orthogonal quantum states, we can always define a projective measurement that enables us to distinguish the states. Consider a set $\{|\psi_i\rangle\}$ of orthonormal states. Define measurement operators M_i via

$$M_i = |\psi_i\rangle\langle \psi_i|, \tag{2.91}$$

one for each possible index i, and one additional measurement operator

$$M_0 = \mathbb{I} - \sum_{i \neq 0} |\psi_i\rangle\langle \psi_i|. \tag{2.92}$$

These operators satisfy the completeness relation and, if the state $|\psi_i\rangle$ is prepared, then the probability of getting outcome i is

$$P_{\psi_i}(i) = \langle\psi_i|M_i|\psi_i\rangle = 1. \tag{2.93}$$

If the states are not orthonormal, the situation is different. In this case, there is no quantum measurement that can reliably distinguish the states. This can be seen by the following argument: Suppose we want to determine which state $|\psi_i\rangle$ from a set of non-orthogonal states $\{|\psi_i\rangle\}$ was given to us. For this purpose, we use a measurement defined by operators M_j with outcomes j. Depending on the measurement outcome we try to guess the index i of the quantum state we have measured, using some rule $f(\cdot)$, i.e., $i = f(j)$. For simplicity, consider the case where we only want to distinguish two states, $|\psi_1\rangle$ and $|\psi_2\rangle$. Suppose now that j is some measurement outcome such that $f(j) = 1$, hence we guess that the state that was given to us is $|\psi_1\rangle$ when observing the outcome j. However, because the states are non-orthogonal, the state $|\psi_2\rangle$ has a component parallel to $|\psi_1\rangle$, hence there is a finite probability that we obtain the outcome j when $|\psi_2\rangle$ is prepared. Therefore, we sometimes make an error when trying to identify the state that was prepared.

Sometimes when performing a quantum mechanical experiment, we are not interested in the state of the system after the measurement but only in the measurement outcome. If this is the case, the mathematical formalism of a *positive operator-valued measure* (POVM) is especially helpful. It can be derived as follows: Suppose we have a measurement described by a set of measurement operators $\{M_x\}$. If the system is in the state $|\psi\rangle$, the probability of measuring the outcome x is given by $P_\psi(x) = \langle\psi|M_x^\dagger M_x|\psi\rangle$. We can define a new operator E_x via

$$E_x = M_x^\dagger M_x. \tag{2.94}$$

This is a positive operator that fulfils the condition $\sum_x E_x = \mathbb{I}$. From the definition it follows that

$$P_\psi(x) = \langle\psi|E_x|\psi\rangle, \tag{2.95}$$

hence the set of operators $\{E_x\}$ is sufficient to describe the measurement statistics of the system. The formal definition of a POVM is the following:

Definition 2.37 A POVM with finite outcome set X is a collection E of operators E_x, indexed by $x \in X$, that satisfy the following properties:

$$\forall x \in X: \ E_x \geq 0, \qquad \sum_{x \in X} E_x = \mathbb{I}. \tag{2.96}$$

If the state of the system is given by a density operator ρ, the probability of obtaining an outcome x is given by

$$P_\rho(x) = \text{Tr}\,(\rho E_x)\,. \tag{2.97}$$

The expectation value of the outcomes is given by

$$\sum_{x \in X} x\,\text{Tr}\,(\rho E_x) \equiv \text{Tr}\,(\rho E)\,, \tag{2.98}$$

where $E = \sum_x x E_x$. This can also be interpreted as the expectation value of the observable E:

$$\langle E \rangle_\rho = \text{Tr}\,(\rho E)\,. \tag{2.99}$$

To illustrate that POVMs are a more general concept than projective measurements, let us have a look at the following situation. We have seen above that two non-orthogonal states cannot be distinguished with certainty. Using a projective measurement, there is a certain probability to misidentify the state. However, we can construct a POVM with *three* measurements that distinguishes the two states *some of the time*, but never makes a wrong identification. Consider a scenario where Alice chooses between two states, $|\psi_0\rangle = |0\rangle$ and $|\psi_1\rangle = \frac{|0\rangle+|1\rangle}{\sqrt{2}}$. The POVM elements are

$$E_0 = \frac{\sqrt{2}}{1+\sqrt{2}}|1\rangle\langle 1| \tag{2.100}$$

$$E_1 = \frac{\sqrt{2}}{1+\sqrt{2}}\frac{(|0\rangle - |1\rangle)\,(\langle 0| - \langle 1|)}{2} \tag{2.101}$$

$$E_2 = \mathbb{I} - E_0 - E_1\,. \tag{2.102}$$

One can quickly verify that these are positive operators with $\sum_i E_i = \mathbb{I}$. If the measurement outcome is 0, the state that was measured can only be $|\psi_1\rangle$ since $\langle\psi_0|E_0|\psi_0\rangle = 0$. If the measurement outcome is 1, the state must have been $|\psi_0\rangle$ since $\langle\psi_1|E_1|\psi_1\rangle = 0$. However, if the measurement outcome is 2, one cannot make a statement about the state since

$$\langle\psi_0|E_2|\psi_0\rangle = \langle\psi_1|E_2|\psi_1\rangle = \frac{1}{2}\,. \tag{2.103}$$

Hence, this POVM correctly identifies the state if the outcome is 0 or 1, but if the outcome is 2 we do not gain any information about the state.

Exercise 2.38 Verify Eq. (2.103).

2.3   Composite Systems and Entanglement

In our study of quantum key distribution protocols, we will often have to deal with
composite systems, for example, the composite system that is composed of Alice's
system and Bob's system, or the one composed of Alice's system and Eve's system.
To talk about these kinds of systems we use the *tensor product*.

The simplest case is to consider two independent quantum systems, A and B
(for example, Alice's lab and Bob's lab), where we run completely independent
experiments and view them as parts of a combined system. The state of the
combined system is then simply the tensor product of the local states: $\rho_{AB} =
\rho_A \otimes \rho_B$ (and similar for the quantum channels and measurements). This is depicted
in Fig. 2.5. Of course, these are not the only possible composite systems. In this
section we describe the mathematical language of composite systems and study
phenomena that arise within these kinds of systems such as entanglement.

2.3.1   States in Composite Systems

As described above, we always associate a Hilbert space to a quantum mechanical
system. Suppose we have a composite system that consists of two subsystems, A and
B (for example, Alice's and Bob's system). Then, the Hilbert space of the composite
system is given by $\mathcal{H}_A \otimes \mathcal{H}_B$.

As an example, consider a composite system of two qubits. As we have seen in
Sect. 2.2.1, each system can be described by a Hilbert space which has as a basis the
computational basis, i.e., $\{|0\rangle, |1\rangle\}$. The basis of the composite system is then given
by $\{|00\rangle, |01\rangle, |10\rangle, |11\rangle\}$. Hence, every possible linear combination of states from
this set is a possible two-qubit state:

$$|\psi\rangle = \alpha|00\rangle + \beta|01\rangle + \gamma|10\rangle + \delta|11\rangle. \tag{2.104}$$

Analogously to the single qubit case, the unit-norm condition $|\alpha|^2 + |\beta|^2 + |\gamma|^2 +
|\delta|^2 = 1$ must hold for the two-qubit state to represent a physical quantum state.

Fig. 2.5 Simple composite
system. We can combine a
product preparation with a
product quantum channel and
a product measurement,
which yields a simple
example of a composite
system

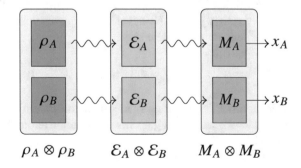

Since the single qubit Hilbert spaces have dimension two, the composite Hilbert space has dimension four (according to (2.16)). Hence, a vector representation of the two-qubit basis can be obtained by taking the tensor product (2.18) of the single-qubit bases given in (2.22). This yields

$$|00\rangle = \begin{pmatrix} 1 \\ 0 \\ 0 \\ 0 \end{pmatrix}, \quad |01\rangle = \begin{pmatrix} 0 \\ 1 \\ 0 \\ 0 \end{pmatrix}, \quad |10\rangle = \begin{pmatrix} 0 \\ 0 \\ 1 \\ 0 \end{pmatrix}, \quad |11\rangle = \begin{pmatrix} 0 \\ 0 \\ 0 \\ 1 \end{pmatrix}. \tag{2.105}$$

The vector representation of the state in (2.104) is then

$$|\psi\rangle = \begin{pmatrix} \alpha \\ \beta \\ \gamma \\ \delta \end{pmatrix}. \tag{2.106}$$

Another basis for the two-qubit state space which is often used in quantum key distribution settings is the so-called *Bell basis*, which consists of the four Bell states:

$$|\Phi^+\rangle = \frac{|00\rangle + |11\rangle}{\sqrt{2}}, \qquad |\Phi^-\rangle = \frac{|00\rangle - |11\rangle}{\sqrt{2}} \tag{2.107}$$

$$|\Psi^+\rangle = \frac{|10\rangle + |01\rangle}{\sqrt{2}}, \qquad |\Psi^-\rangle = \frac{|10\rangle - |01\rangle}{\sqrt{2}}. \tag{2.108}$$

Exercise 2.39 Verify that the four Bell states form an orthonormal basis of a two-qubit system.

Measuring one of the bits of a Bell state automatically determines the second bit. Consider the state $|\Phi^+\rangle_{AB} \in \mathcal{H}_A \otimes \mathcal{H}_B$ and suppose we make a measurement of the first qubit in the computational basis while leaving the second qubit unchanged. The corresponding measurement operators are

$$M_0 = |0\rangle_A \langle 0| \otimes \mathbb{I}_B, \qquad M_1 = |1\rangle_A \langle 1| \otimes \mathbb{I}_B. \tag{2.109}$$

The probability of getting the outcome 0 is

$$P_{\Phi^+}(0) = {}_{AB}\langle \Phi^+|M_0|\Phi^+\rangle_{AB} \tag{2.110}$$

$$= \frac{1}{2}\Big({}_A\langle 0|_B\langle 0| + {}_A\langle 1|_B\langle 1| \Big)|0\rangle_A\langle 0| \otimes \mathbb{I}_B\Big(|0\rangle_A|0\rangle_B + |1\rangle_A|1\rangle_B\Big) \tag{2.111}$$

$$= \frac{1}{2}, \tag{2.112}$$

and the probability of getting outcome 1 is $P_{\Phi^+}(1) = \frac{1}{2}$. The state after the measurement, depending on the measurement outcome, is given by

$$|\psi_0'\rangle = \sqrt{2}\big(|0\rangle_A\langle 0| \otimes \mathbb{I}_B\big)|\Phi^+\rangle_{AB} = |0\rangle_A|0\rangle_B \tag{2.113}$$

$$|\psi_1'\rangle = \sqrt{2}\big(|1\rangle_A\langle 1| \otimes \mathbb{I}_B\big)|\Phi^+\rangle_{AB} = |1\rangle_A|1\rangle_B. \tag{2.114}$$

Therefore, the state of the second qubit is determined even though the measurement has only taken place locally on system A.

Mixed states, which are represented by density matrices, can be composed in an analogous way to pure states. Suppose we have states $\rho_A \in \mathcal{B}(\mathcal{H}_A)$ and $\rho_B \in \mathcal{B}(\mathcal{H}_B)$. The composite state is then simply $\rho_A \otimes \rho_B \in \mathcal{B}(\mathcal{H}_A \otimes \mathcal{H}_B)$. This is also consistent with the interpretation of the density matrix as a description of a system whose state is not precisely know. Suppose we have an ensemble $\{p_i, |\phi_i\rangle\}_i$ for system A and $\{q_j, |\psi_j\rangle\}_j$ for system B. Then, the ensemble for the composite system is $\{p_i q_j, |\phi_i\rangle \otimes |\psi_j\rangle\}_{i,j}$ and the corresponding density operator is given by

$$\rho_{AB} = \sum_{i,j} p_i p_j \left(|\phi_i\rangle \otimes |\psi_j\rangle\right)\left(\langle\phi_i| \otimes \langle\psi_j|\right) \tag{2.115}$$

$$= \sum_{i,j} p_i q_j |\phi_i\rangle\langle\phi_i| \otimes |\psi_j\rangle\langle\psi_j| \tag{2.116}$$

$$= \sum_i p_i |\phi_i\rangle\langle\phi_i| \sum_j q_j |\psi_j\rangle\langle\psi_j| \tag{2.117}$$

$$= \rho_A \otimes \rho_B. \tag{2.118}$$

2.3.2 Entanglement

Of course, product states are not the only possible states in the composite Hilbert space. There are other states that do not exhibit such a product form, and they have astounding properties. If two (or more) spatially separated parties share a quantum state, the state can exhibit quantum correlations. This phenomenon, called *entanglement*, is unique to quantum theory and is used as a resource in quantum cryptography and quantum computing, for example.

Let us first study this phenomenon for pure states: Suppose two parties, Alice and Bob, share the two-qubit quantum state

$$|0\rangle_A|0\rangle_B, \tag{2.119}$$

where Alice holds the qubit in system A and Bob the one in system B. Alice knows exactly that her qubit is in the state $|0\rangle_A$ and Bob can also definitely say that his qubit is in the state $|0\rangle_A$, so there is no strange correlation happening here.

In contrast to this, consider the composite quantum state

$$|\Phi^+\rangle_{AB} = \frac{1}{\sqrt{2}}\left(|0\rangle_A|0\rangle_B + |1\rangle_A|1\rangle_B\right). \tag{2.120}$$

As before, Alice possesses the qubit in system A and Bob the one in system B. But in contrast to the previous case, Alice cannot determine the individual state of her qubit from the description in (2.120), and neither can Bob. The above state is a superposition of the two-qubit states $|0\rangle_A|0\rangle_B$ and $|1\rangle_A|1\rangle_B$, and it is not possible to determine the individual states that Alice and Bob hold, in the sense that we cannot describe the state $|\Phi^+\rangle$ as a product state of the form $|\phi\rangle_A \otimes |\psi\rangle_B$. We say that the state is *entangled*. This yields the following definition of entangled states:

Definition 2.40 A pure bipartite state $|\psi\rangle_{AB}$ is *entangled* if it cannot be written as a product state $|\phi\rangle_A \otimes |\eta\rangle_B$ for any choice of states $|\phi\rangle_A$ and $|\eta\rangle_B$. Otherwise, it is called *separable*.

Definition 2.41 Consider a bipartite system $\mathcal{H}_A \otimes \mathcal{H}_B$ with $\dim(\mathcal{H}_A) = \dim(H_B) = d$ and orthonormal bases $\{|i\rangle_A\}$ and $\{|i\rangle_B\}$. The *maximally entangled state* is then defined as

$$|\Omega\rangle = \frac{1}{\sqrt{d}} \sum_{i=1}^{d} |ii\rangle. \tag{2.121}$$

Exercise 2.42 Decide for each of the following states whether it is separable or entangled:

1. $|\psi_1\rangle = |11\rangle$.
2. $|\psi_2\rangle = \frac{1}{2}\left(|00\rangle - |01\rangle + |10\rangle - |11\rangle\right)$.
3. $|\psi_3\rangle = \frac{1}{\sqrt{2}}\left(|10\rangle + |01\rangle\right)$.

For mixed states, we define entanglement in a similar way:

Definition 2.43 A bipartite state $\rho_{AB} \in \mathcal{B}(\mathcal{H}_A \otimes \mathcal{H}_B)$ is called *separable* if and only if it can be written as a linear convex combination of tensor products of density matrices

$$\rho_{AB} = \sum_{x} p_x \sigma_A^x \otimes \eta_B^x, \tag{2.122}$$

for some probability distribution p_x and states $\sigma_A^x \in \mathcal{B}(\mathcal{H}_A)$ and $\eta_B^x \in \mathcal{B}(\mathcal{H}_B)$. Otherwise, ρ_{AB} is called *entangled*.

Deciding whether a given state is entangled or not is not always obvious. Fortunately, we have a powerful tool to analyse bipartite pure states, namely the *Schmidt decomposition*:

Theorem 2.44 (Schmidt Decomposition) *Let $|\psi\rangle$ be a pure state in the Hilbert space $\mathcal{H}_A \otimes H_B$. Then,*

$$|\psi\rangle = \sum_{i=1}^{d} \lambda_i |i\rangle_A |i\rangle_B, \qquad (2.123)$$

where the amplitudes λ_i are real, strictly positive, and $\sum_i \lambda_i^2 = 1$. The states $\{|i\rangle_A\}_i$ form an orthonormal basis for the system A and the states $\{|i\rangle_B\}_i$ form an orthonormal basis for the system B. The amplitudes λ_i are called Schmidt coefficients and the Schmidt rank d is equal to the number of Schmidt coefficients λ_i and satisfies

$$d \leq \min\{\dim(\mathcal{H}_A), \dim(\mathcal{H}_B)\}. \qquad (2.124)$$

Proof The key ingredient for the proof of the Schmidt decomposition is the singular value decomposition. We first give the proof for the case that the Hilbert spaces have the same dimension, i.e., $d = \dim(\mathcal{H}_A) = \dim(\mathcal{H}_B)$. Consider an arbitrary bipartite pure state $|\psi\rangle$, which can be written as

$$|\psi\rangle = \sum_{j,k} c_{jk} |j\rangle_A |k\rangle_B, \qquad (2.125)$$

with orthonormal bases $\{|j\rangle_A\}$ and $\{|k\rangle_B\}$ on system A and B, respectively, and some amplitudes c_{jk}. The amplitudes c_{jk} form a matrix $C = (c_{jk})$, which admits a singular value decomposition

$$C = U \Lambda V, \qquad (2.126)$$

where U and V are $d \times d$ unitaries and Λ is a diagonal matrix with non-negative, real numbers λ_i on the diagonal. Writing the matrix elements of U as u_{ji} and the matrix elements of V as v_{ik}, the above matrix equation is equivalent to the following set of equations:

$$c_{jk} = \sum_i u_{ji} \lambda_i v_{ik}. \qquad (2.127)$$

Note that we only need one index for the elements of the matrix Λ since it only has entries on the diagonal. The state $|\psi\rangle$ can then be written as

$$|\psi\rangle = \sum_{j,k} \left(\sum_i u_{ji} \lambda_i v_{ik} \right) |j\rangle_A |k\rangle_B \tag{2.128}$$

$$= \sum_i \lambda_i \underbrace{\left(\sum_j u_{ji} |j\rangle_A \right)}_{\equiv |i\rangle_A} \otimes \underbrace{\left(\sum_k v_{ik} |k\rangle_B \right)}_{\equiv |i\rangle_B} \tag{2.129}$$

$$= \sum_i \lambda_i |i\rangle_A |i\rangle_B, \tag{2.130}$$

where the sets $\{|i\rangle_A\}$ and $\{|i\rangle_B\}$ form orthonormal bases for system A and B, respectively. If $\dim(\mathcal{H}_A) \neq \dim(\mathcal{H}_B)$ the proof works analogously. $\qquad\square$

Exercise 2.45 Verify that the set of states $\{|i\rangle_A\}$ with $|i\rangle_A = \sum_j u_{ji} |j\rangle_A$ is an orthonormal basis of system A. *Hint: Use that U is a unitary matrix.*

Note that the Schmidt decomposition not only applies to bipartite systems, but to any multipartite system where we can make a bipartite cut of the systems.

The statement of Theorem 2.44 is practical for several reasons: Firstly, once we know the Schmidt decomposition of a state, we can immediately say whether it is entangled or not. If a state $|\psi\rangle$ is entangled, then its Schmidt decomposition has more than one term. Differently stated, a state is entangled if and only if its Schmidt rank is strictly greater than one. The second reason is the following: Suppose the Hilbert space of Alice $\mathcal{H}_A$ is a qubit Hilbert space, i.e., it is two-dimensional, while Bob's Hilbert space $\mathcal{H}_B$ is of dimension ten million (or some other very large number). The statement of the theorem is then that, although Bob's Hilbert space is extremely large, it is always possible to find a two-dimensional subspace of $\mathcal{H}_B$ which, along with $\mathcal{H}_A$, suffices to represent a pure state in the composite system $\mathcal{H}_A \otimes \mathcal{H}_B$. There is another important consequence of the Schmidt decomposition, but to understand it we first need to introduce the *partial trace operation*.

We have seen above that the tensor product provides a mathematical tool to prepare the state of a composite system given the states of the individual subsystems. Suppose we are given the density operator ρ_{AB} of the composite system $A \otimes B$. Can we make a statement about the individual density operators ρ_A and ρ_B of the subsystems A and B, i.e., a *local* density operator? Recall that the density operator description provides a mathematical representation of the state of the system that allows us to compute the probabilities resulting from a physical measurement. The general method to determine a local density operator which describes the state of

Alice's system if Bob's system is inaccessible to her is to take the *partial trace* of the composite state:

Definition 2.46 Let ρ_{AB} be a density operator acting on the bipartite Hilbert space $\mathcal{H}_A \otimes \mathcal{H}_B$ and let $\{|i\rangle_B\}_i$ be an orthonormal basis for $\mathcal{H}_B$. The partial trace over the Hilbert space $\mathcal{H}_B$ is then defined as follows:

$$\mathrm{Tr}_B\left(\rho_{AB}\right) = \sum_i \left(\mathbb{I}_A \otimes {}_B\langle i|\right) \rho_{AB} \left(\mathbb{I}_A \otimes |i\rangle_B\right). \tag{2.131}$$

For simplicity, we usually suppress the identity operator $\mathbb{I}_A$ and write (2.131) as

$$\mathrm{Tr}_B\left(\rho_{AB}\right) = \sum_i {}_B\langle i| \rho_{AB} |i\rangle_B. \tag{2.132}$$

The state $\rho_A = \mathrm{Tr}_B\left(\rho_{AB}\right)$ is also called the *reduced state* or *marginal* of system A.

Exercise 2.47 Suppose we have a bipartite system in the state $\rho_{AB} \in \mathcal{B}(\mathcal{H}_A \otimes \mathcal{H}_B)$. Show that the reduced state $\rho_A = \mathrm{Tr}_B\left(\rho_{AB}\right)$ is a valid density operator, i.e., that it is positive semi-definite and has trace equal to one.

Exercise 2.48 Show that applying the two partial traces in any order on a bipartite system is equivalent to taking the full trace:

$$\mathrm{Tr}\left(\rho_{AB}\right) = \mathrm{Tr}_A\left(\mathrm{Tr}_B\left(\rho_{AB}\right)\right) = \mathrm{Tr}_B\left(\mathrm{Tr}_A\left(\rho_{AB}\right)\right). \tag{2.133}$$

Given the definition of the partial trace, we can now understand another important consequence of the Schmidt decomposition: Consider a pure state of a composite system, $|\psi\rangle_{AB}$. Using the Schmidt decomposition of the state, it is easy to see that the marginal states of $\rho_{AB} = |\psi\rangle_{AB}\langle\psi|$ are

$$\rho_A = \mathrm{Tr}_B\left(|\psi\rangle_{AB}\langle\psi|\right) = \sum_i \lambda_i^2 |i\rangle_A\langle i|, \tag{2.134}$$

$$\rho_B = \mathrm{Tr}_A\left(|\psi\rangle_{AB}\langle\psi|\right) = \sum_i \lambda_i^2 |i\rangle_B\langle i|, \tag{2.135}$$

so the eigenvalues of ρ_A and ρ_B are identical, namely λ_i^2 for both density operators. Since many properties of quantum systems (especially when it comes to entropies) are determined by the eigenvalues of the density operator, this is extremely practical.

Recall the state $|\Phi^+\rangle_{AB}$ we have introduced above. Using the Schmidt decomposition, we now see the reason why it is difficult to say which states the individual qubits are in. Since $\{|0\rangle, |1\rangle\}$ is an orthonormal basis for the qubit state space, it is straightforward to write down the Schmidt decomposition of $|\Phi^+\rangle_{AB}$:

$$|\Phi^+\rangle_{AB} = \frac{1}{\sqrt{2}} \left(|0\rangle_A |0\rangle_B\right) + \frac{1}{\sqrt{2}} \left(|1\rangle_A |1\rangle_B\right). \tag{2.136}$$

The formula above shows that we have two Schmidt coefficients, namely $\lambda_1 = \lambda_2 = \frac{1}{\sqrt{2}}$, which fulfil the requirement that $\lambda_1^2 + \lambda_2^2 = 1$. Hence, the Schmidt rank of the state is $d = 2$ and therefore, $|\Phi^+\rangle_{AB}$ is an entangled state. This is the reason why we cannot easily assign local states to the involved qubits.

Even though the state $|\Phi^+\rangle_{AB}$ cannot be written as a product of local states, we can still describe the state of Alice's system given that Bob's system is inaccessible to her using the partial trace operation. To take the partial trace over Bob's system we need an orthonormal basis for his Hilbert space. Since we are considering a two-qubit state that Alice and Bob share, a suitable orthonormal basis is $\{|0\rangle, |1\rangle\}$. Hence, we can calculate the reduced state for Alice's system in the following way:

$$\rho_A = \mathrm{Tr}_B \left(|\Phi^+\rangle_{AB}\langle\Phi^+|\right)$$

$$= \frac{1}{2}\Big(\left(\mathbb{I}_A \otimes {}_B\langle 0|\right)\left(|0\rangle_A|0\rangle_B + |1\rangle_A|1\rangle_B\right)\left({}_A\langle 0|{}_B\langle 0| + {}_A\langle 1|{}_B\langle 1|\right)\left(\mathbb{I}_A \otimes |0\rangle_B\right)$$

$$+ \left(\mathbb{I}_A \otimes {}_B\langle 1|\right)\left(|0\rangle_A|0\rangle_B + |1\rangle_A|1\rangle_B\right)\left({}_A\langle 0|{}_B\langle 0| + {}_A\langle 1|{}_B\langle 1|\right)\left(\mathbb{I}_A \otimes |1\rangle_B\right)\Big)$$

$$= \frac{1}{2}\Big(|0\rangle_A\langle 0| + |1\rangle_A\langle 1|\Big)$$

$$= \pi_A,$$

where π_A is the maximally mixed state introduced in Example 2.18. With a similar calculation we find that Bob's reduced density operator is $\rho_B = \pi_B$. Can we conclude from this calculation that $\rho_{AB} = \pi_A \otimes \pi_B$? Certainly not! It is easy to see that $|\Phi^+\rangle_{AB}\langle\Phi^+| \neq \pi_A \otimes \pi_B$ by simply writing down the vector representation of the states.

Exercise 2.49 On an even more fundamental level, the global state $|\Phi^+\rangle_{AB}\langle\Phi^+|$ and the state $\pi_A \otimes \pi_B$ give different predictions for global measurements, i.e., measurements on the composite system $A \otimes B$. Consider the so-called *parity measurement* which is given by the operators

$$\Pi_{\mathrm{even}} = |00\rangle_{AB}\langle 00| + |11\rangle_{AB}\langle 11|, \tag{2.137}$$

$$\Pi_{\mathrm{odd}} = |01\rangle_{AB}\langle 01| + |10\rangle_{AB}\langle 10|. \tag{2.138}$$

1. Show that the probability of getting an even parity as a result of the parity measurement for the state $|\Phi^+\rangle_{AB}\langle\Phi^+|$ is 1, i.e., $P_{\Phi^+}(\text{even}) = 1$.
2. Show that the probability of getting an even parity as a result of the parity measurement for the state $\pi_A \otimes \pi_B$ is $\frac{1}{2}$, i.e., $P_{\pi_A \otimes \pi_B}(\text{even}) = \frac{1}{2}$.
3. Verify that $P_{\Phi^+}(\text{odd}) = 0$ and $P_{\pi_A \otimes \pi_B}(\text{odd}) = \frac{1}{2}$.

This shows that the states give different predictions for the outcome of a measurement and therefore cannot describe the same quantum system. Hence, it really is impossible to find local states of the state $|\Phi^+\rangle_{AB}\langle\Phi^+|$ for Alice's system and Bob's system, respectively, that describe the probabilities of outcomes of global measurements correctly. Still, it is possible to find local states that describe Alice's knowledge of her local system if she does not have access to Bob's system.

Exercise 2.50 Show that all of the Bell states are entangled by calculating the Schmidt rank for the remaining three Bell states.

2.3.3   Quantum–Classical Ensemble

With the tools that we have developed we can describe not only multipartite systems where each subsystem is quantum but also hybrid systems that consist of both classical and quantum subsystems.

For the description of a quantum–classical hybrid state, we first have to define an ensemble of a purely classical system. Since the density operator formalism holds for general states, this should also include classical states. In the same way that the density matrix of a quantum state can be interpreted as the description of a system whose precise state is not known, we now need a notion of classical randomness. This is provided by a *random variable* Z: Suppose we have a set of classical values $\mathcal{Z}$ whose entries $z \in \mathcal{Z}$ (also called *realizations* of the random variable Z) are distributed according to the probability distribution $p_Z(z)$. We can represent these classical values by orthonormal states[4] $|z\rangle$ in some Hilbert space $\mathcal{H}_Z$. The density operator ρ_Z that corresponds to the classical ensemble $\{p_Z(z), |z\rangle\langle z|\}$ is then

$$\rho_Z = \sum_{z \in \mathcal{Z}} p_Z(z)|z\rangle\langle z|. \tag{2.139}$$

We can now extend this consideration to quantum–classical hybrid systems. Consider the tensor product Hilbert space $\mathcal{H}_A \otimes \mathcal{H}_Z$, where $\mathcal{H}_Z$ is a Hilbert system of states that represent classical values distributed by the probability distribution $p_Z(z)$ and $\mathcal{H}_A$ is the Hilbert space of a quantum system whose states ρ_A^z depend on

---

[4]It is crucial that the states are orthonormal, because it ensures that they are perfectly distinguishable, which is always the case for classical values.

the classical value z. The corresponding ensemble is

$$\left\{p_Z(z), \rho_A^z \otimes |z\rangle z \langle z|\right\}_{z\in\mathcal{Z}}. \tag{2.140}$$

The density operator that corresponds to this quantum–classical ensemble is then

$$\rho_{AZ} = \sum_{z\in\mathcal{Z}} p_Z(z)\rho_A^z \otimes |z\rangle z \langle z|. \tag{2.141}$$

This is a special kind of separable state of the systems A and Z, where the individual states of the system Z are perfectly distinguishable and therefore classical.

2.3.4 Evolution of Composite Systems

If you have a state on a composite system of two individual systems, e.g., the tensor product Hilbert space $\mathcal{H}_A \otimes \mathcal{H}_B$, the quantum evolution is a linear, completely positive, trace-preserving map, according to Definition 2.25. The only difference to the previous case is that the map now goes between tensor product Hilbert spaces, i.e., we have a map $\mathcal{E}_{AB} : \mathcal{B}(\mathcal{H}_A \otimes \mathcal{H}_B) \rightarrow \mathcal{B}(\mathcal{H}'_A \otimes \mathcal{H}'_B)$. In a bipartite quantum system, there are basically three kinds of evolution the state can undergo (as depicted in Fig. 2.6): The evolution only affects system A while system B is left unchanged, it only affects system B while system A is left unchanged, or it affects both systems.

Let us have a look at the special cases depicted in Fig. 2.6a and b. Here, the quantum channel acts on one of the subsystems and leaves the other subsystem unchanged, i.e., it acts as the identity on this subsystem. Suppose we have a state $\rho_{AB} \in \mathcal{B}(\mathcal{H}_A \otimes \mathcal{H}_B)$ and a quantum channel $\mathcal{E}_A$ that only acts on system A (which corresponds to Fig. 2.6a). The complete evolution is then given by

$$\rho'_{AB} = (\mathcal{E}_A \otimes \mathbb{I}_B)(\rho_{AB}). \tag{2.142}$$

This can be analogously defined for a quantum channel that only acts on Bob's subsystem (as depicted in Fig. 2.6b).

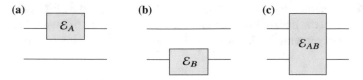

Fig. 2.6 The possible evolutions of a bipartite system. (**a**) The evolution happens only on subsystem A and leaves subsystem B unchanged. (**b**) The evolution happens only on subsystem B and leaves subsystem A unchanged. (**c**) The whole composite system $A \otimes B$ undergoes an evolution (the identity evolution, where nothing happens to the whole system, is a special case of all three cases)

By using such a local quantum evolution, it is possible to change between any of the four Bell states. Suppose Alice and Bob share one of the Bell states. Applying the Pauli-X or the Pauli-Z operator locally to Alice's part of the system has the following effect:

$$(X_A \otimes \mathbb{I}_B)|\Phi^\pm\rangle_{AB} = |\Psi^\pm\rangle_{AB}, \tag{2.143}$$

$$(X_A \otimes \mathbb{I}_B)|\Psi^\pm\rangle_{AB} = |\Phi^\pm\rangle_{AB}, \tag{2.144}$$

$$(Z_A \otimes \mathbb{I}_B)|\Phi^\pm\rangle_{AB} = |\Phi^\mp\rangle_{AB}, \tag{2.145}$$

$$(Z_A \otimes \mathbb{I}_B)|\Psi^\pm\rangle_{AB} = |\Psi^\mp\rangle_{AB}. \tag{2.146}$$

Exercise 2.51 Verify Eqs. (2.143)–(2.146).

Exercise 2.52 What is the effect of applying the Pauli-Y operator to the first qubit of the four Bell states?

As a second example, one can view the partial trace operation from Definition 2.46 as a quantum channel that only acts on one of the subsystems. Suppose we have a state $\rho_{AB} \in \mathcal{B}(\mathcal{H}_A \otimes \mathcal{H}_B)$ and want to take the partial trace over subsystem B, i.e., $\mathcal{E}_B = \mathrm{Tr}_B$:

$$\mathrm{Tr}_B(\rho_{AB}) = (\mathbb{I}_A \otimes \mathrm{Tr}_B)(\rho_{AB}) = \sum_i (\mathbb{I}_A \otimes {}_B\langle e_i|) \, \rho_{AB} \, (\mathbb{I}_A \otimes |e_i\rangle_B). \tag{2.147}$$

This channel is called the *discarding channel*. The Kraus operators of the channel $\mathcal{E}_{AB} = \mathbb{I}_A \otimes \mathrm{Tr}_B$ are given by the set $\{\mathbb{I}_A \otimes {}_B\langle e_i|\}_i$, where $\{|e_i\rangle\}_i$ is an orthonormal basis of the Hilbert space $\mathcal{H}_B$.

2.4   The No-Cloning Theorem

The no-cloning theorem, which was originally formulated in 1982 [6], states that it is impossible to build a *universal copier* of quantum states, i.e., a machine that can perfectly clone an unknown quantum state as depicted in Fig. 2.7. This is a consequence of the linearity of quantum theory and it lies at the heart of the security of quantum cryptography: It prevents the adversary from simply cloning all the

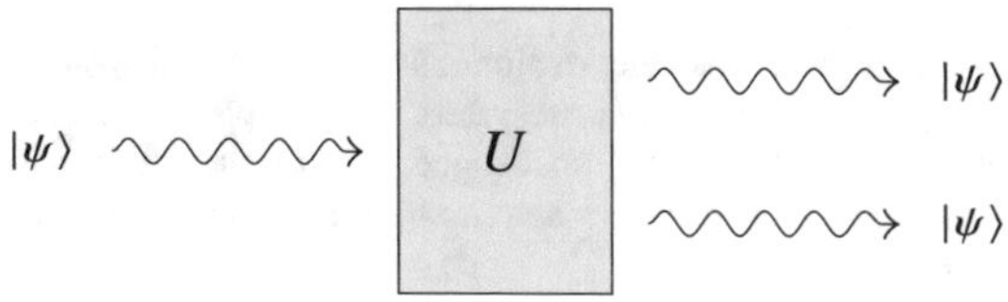

Fig. 2.7 A universal copier. This device, described by a unitary operator U, copies an arbitrary, unknown quantum state $|\psi\rangle$, such that we get two perfect copies of the state

states Alice sends to Bob, which would allow her to gain information without being detected.

The theorem can be proved via contradiction: Suppose we have a unitary operator U that acts on two qubits as a universal copier of quantum information. More precisely, if we input an arbitrary qubit state $|\psi\rangle = \alpha|0\rangle + \beta|1\rangle$ as the first qubit and an ancilla qubit in the state $|0\rangle^5$ as the second qubit, the operator U should write the state of the first qubit to the second qubit slot:

$$U|\psi\rangle|0\rangle = |\psi\rangle|\psi\rangle \tag{2.148}$$

$$= (\alpha|0\rangle + \beta|1\rangle)(\alpha|0\rangle + \beta|1\rangle) \tag{2.149}$$

$$= \alpha^2|0\rangle|0\rangle + \alpha\beta|0\rangle|1\rangle + \beta\alpha|1\rangle|0\rangle + \beta^2|1\rangle|1\rangle. \tag{2.150}$$

Since the copier is universal, it can copy an arbitrary state. In particular, it can copy the states $|0\rangle$ and $|1\rangle$:

$$U|0\rangle|0\rangle = |0\rangle|0\rangle, \qquad U|1\rangle|0\rangle = |1\rangle|1\rangle. \tag{2.151}$$

We now use the fact that quantum theory is linear: This implies that the unitary operator U acts on a superposition of the two states as follows:

$$U(\alpha|0\rangle + \beta|1\rangle)|0\rangle = \alpha U|0\rangle|0\rangle + \beta U|1\rangle|0\rangle = \alpha|0\rangle|0\rangle + \beta|1\rangle|1\rangle. \tag{2.152}$$

Comparing the result of (2.152) with (2.150) yields a contradiction: The two expressions do not have to be equal for all α and β. There exist choices of α and β for which

$$\alpha^2|0\rangle|0\rangle + \alpha\beta|0\rangle|1\rangle + \beta\alpha|1\rangle|0\rangle + \beta^2|1\rangle|1\rangle \neq \alpha|0\rangle|0\rangle + \beta|1\rangle|1\rangle. \tag{2.153}$$

Hence, the linearity of quantum theory makes it impossible to build a universal copier of quantum states. Note that although the expressions in (2.153) are different in general, there exist special cases for which they are equal, namely if $\alpha = 1, \beta = 0$ or $\alpha = 0, \beta = 1$. These cases correspond to classical states, i.e., we are able to copy unknown classical states in the basis $\{|0\rangle, |1\rangle\}$ (or any other orthonormal basis, for that matter).

The no-cloning theorem reveals a fundamental difference between classical and quantum information theory. This is crucial for the security of quantum cryptography, since it prevents the adversary from perfectly copying the states that are send without being detected.

---

[5]The initial state of the ancilla qubit does not actually matter. We could write any state as a placeholder.

2.5   Purification

In this section, we introduce the concept of *purification*, which allows us to view
noise in quantum systems as the result of the system being entangled with another
system which we do not have access to.

Consider a density operator $\rho_A \in \mathcal{B}(\mathcal{H}_A)$ with spectral decomposition

$$\rho_A = \sum_i p_i |\varphi_i\rangle_A \langle\varphi_i|. \tag{2.154}$$

From the spectral decomposition we can deduce the corresponding ensemble
$\{p_i, |\varphi_i\rangle_A\}$. We can now associate a pure state on a larger Hilbert space to this,
in general, mixed state:

Definition 2.53 A purification of $\rho_A \in \mathcal{B}(\mathcal{H}_A)$ is a pure bipartite state $|\psi\rangle_{RA} \in$
$\mathcal{H}_R \otimes \mathcal{H}_A$ on a reference system R and on the original system A. If we trace out
reference system R, the reduced state on system A is equal to ρ_A:

$$\rho_A = \mathrm{Tr}_R\big(|\psi\rangle_{RA}\langle\psi|\big). \tag{2.155}$$

Any density operator ρ_A has a purification $|\psi\rangle_{RA}$ given by

$$|\psi\rangle_{RA} = \sum_i \sqrt{p_i} |\varphi_i\rangle_R |\varphi_i\rangle_A, \tag{2.156}$$

where $\{|\varphi_i\rangle_R\}$ is a set of orthonormal vectors on the reference system R.[6] It can
quickly be checked that this state indeed is a purification of ρ_A by tracing out the
reference system:

$$\mathrm{Tr}_R\big(|\psi\rangle_{RA}\langle\psi|\big) = \mathrm{Tr}_R\left(\sum_{i,j} \sqrt{p_i p_j}\, |\varphi_i\rangle_R |\varphi_i\rangle_A\, {}_R\langle\varphi_j|_A\langle\varphi_j|\right) \tag{2.157}$$

$$= \sum_{i,j} \sqrt{p_i p_j}\, \underbrace{\mathrm{Tr}_R\big(|\varphi_i\rangle_R \langle\varphi_j|\big)}_{=\delta_{ij}}\, |\varphi_i\rangle_A \langle\varphi_j| \tag{2.158}$$

$$= \sum_i p_i |\varphi_i\rangle_A \langle\varphi_i| = \rho_A. \tag{2.159}$$

Exercise 2.54 Show that any of the four Bell states is a purification of the
maximally mixed state on Alice's system $\pi_A = \frac{\mathbb{I}_A}{2}$.

---

[6]Note that there are, in general, infinitely many possible purifications for a given quantum state ρ.

We can also define the notion of an *extension* of a quantum state ρ_A, which is some (not necessarily pure) quantum state $\sigma_{RA} \in \mathcal{B}(\mathcal{H}_R \otimes \mathcal{H}_A)$ such that

$$\rho_A = \text{Tr}_R (\sigma_{RA}) . \tag{2.160}$$

This definition is useful in some cases. However, it is always possible to find a purification of an extension.

2.6 Distance Measures

Throughout our analysis of quantum key distribution protocols, a crucial point is to estimate how much the actual key differs from the optimal one. For this purpose, we need a measure of distance between quantum states. One possible measure can be defined via the *trace norm* (also called L_1-norm or *Schatten 1-norm*):

Definition 2.55 The trace norm of a state $\rho \in \mathcal{B}(\mathcal{H})$ is defined as

$$||\rho||_1 = \text{Tr} (|\rho|) , \tag{2.161}$$

where we define $|\rho| = \sqrt{\rho^\dagger \rho}$ to be the positive square root of $\rho^\dagger \rho$.

This norm induces a natural distance measure for quantum states, called the *trace distance*: The trace distance of two quantum states $\rho, \sigma \in \mathcal{B}(\mathcal{H})$ is defined to be $\frac{1}{2}||\rho - \sigma||_1$. The factor of $\frac{1}{2}$ in front of the norm is chosen because

$$0 \leq ||\rho - \sigma||_1 \leq 2 \tag{2.162}$$

for any two density operators ρ and σ. Hence, dividing this expression by 2 normalizes the trace distance and we have $\frac{1}{2}||\rho - \sigma||_1 \in [0, 1]$.

The trace norm has some important properties: First, it fulfils the triangle inequality: For two operators $\rho, \sigma \in \mathcal{B}(\mathcal{H})$ it holds that

$$||\rho + \sigma||_1 \leq ||\rho||_1 + ||\sigma||_1 . \tag{2.163}$$

Another property of the trace distance is that no quantum operation can ever increase the distance between two quantum states, which is depicted in Fig. 2.8 and formally expressed in the following theorem, whose proof can be found in [3, Thm. 9.2]:

Theorem 2.56 *Suppose* $\mathcal{E} : \mathcal{B}(\mathcal{H}_A) \rightarrow \mathcal{B}(\mathcal{H}_B)$ *is a completely positive, trace-preserving map. Let* $\rho, \sigma \in \mathcal{B}(\mathcal{H}_A)$ *be density operators. Then*

$$||\mathcal{E}(\rho) - \mathcal{E}(\sigma)||_1 \leq ||\rho - \sigma||_1 . \tag{2.164}$$

Fig. 2.8 CPTP maps cause a contraction. A quantum operation $\mathcal{E}$ can never increase the distance between two quantum states $\rho, \sigma \in \mathcal{B}(\mathcal{H})$. The circle represents the space of density operators $\mathcal{B}(\mathcal{H})$

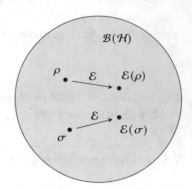

A second important tool to compare two quantum states is the *fidelity*, which measures the *closeness* of two quantum states:

Definition 2.57 For two quantum states $\rho, \sigma \in \mathcal{B}(\mathcal{H})$, the fidelity is defined as

$$F(\rho, \sigma) = \left(\text{Tr} \left(\sqrt{\rho^{\frac{1}{2}} \sigma \rho^{\frac{1}{2}}} \right) \right)^2. \tag{2.165}$$

If one of the quantum states is a pure state, the above formula simplifies to

$$F(|\psi\rangle, \rho) = \left(\text{Tr} \left(\sqrt{|\psi\rangle\langle\psi|\rho|\psi\rangle\langle\psi|} \right) \right)^2 \tag{2.166}$$

$$= \langle\psi|\rho|\psi\rangle\langle\psi|\psi\rangle \tag{2.167}$$

$$= \langle\psi|\rho|\psi\rangle, \tag{2.168}$$

i.e., the fidelity is equal to the overlap between ρ and $|\psi\rangle$. If both states are pure states, the fidelity takes an even simpler form:

$$F(|\psi\rangle, |\phi\rangle) = |\langle\psi|\phi\rangle|^2. \tag{2.169}$$

The term on the right hand side can also be interpreted as the probability that the state $|\psi\rangle$ would pass a test for being the same as $|\phi\rangle$.

The fidelity is sometimes also presented in a slightly different form, namely as the square root of the right hand side in (2.165):

$$F'(\rho, \sigma) = \text{Tr} \left(\sqrt{\rho^{\frac{1}{2}} \sigma \rho^{\frac{1}{2}}} \right) = \sqrt{F(\rho, \sigma)}. \tag{2.170}$$

This definition of fidelity is less common, and to avoid confusion it could be called *square root fidelity*. In some applications the square root fidelity has an easier form than the original definition. However, in these notes we stick to the definition of fidelity given in Definition 2.57.

Strictly speaking, the fidelity is not a metric on the space of density operators since it measures closeness rather than distance. However, it can be used to define a metric: one possibility of turning the fidelity into a proper distance measure is the *purified distance*[7] (see [1]), which we will need later in these notes. It is defined as follows:

Definition 2.58 For two quantum states $\rho, \sigma \in \mathcal{B}(\mathcal{H})$, the purified distance between ρ and σ is defined as

$$P(\rho, \sigma) = \sqrt{1 - F(\rho, \sigma)}. \tag{2.171}$$

Even though the fidelity is not a metric itself, it has many properties that still make it a useful quantity to estimate how close two quantum states are:

1. It is symmetric in its arguments: $F(\rho, \sigma) = F(\sigma, \rho)$.
2. The fidelity is bounded by $0 \leq F(\rho, \sigma) \leq 1$. The lower bound applies in case ρ and σ have orthogonal supports and the upper bound applies if the two states are equal.
3. For a completely positive, trace-preserving map $\mathcal{E}$ and quantum states ρ and σ,

$$F(\mathcal{E}(\rho), \mathcal{E}(\sigma)) \geq F(\rho, \sigma). \tag{2.172}$$

Since the fidelity measures how close two quantum states are, one would intuitively think that a high fidelity of two quantum states implies a small trace distance and vice versa. This intuition is made precise in the following theorem, which establishes several relationships between the fidelity and the trace distance (for a proof of this theorem, see for example [5, Thm. 9.3.1]):

Theorem 2.59 *The following relationship holds for the fidelity and the trace distance between two quantum states $\rho, \sigma \in \mathcal{B}(\mathcal{H})$:*

$$1 - \sqrt{F(\rho, \sigma)} \leq \frac{1}{2}\|\rho - \sigma\|_1 \leq \sqrt{1 - F(\rho, \sigma)}. \tag{2.173}$$

Exercise 2.60 Using Theorem 2.59, show the following statements for two quantum states $\rho, \sigma \in \mathcal{B}(\mathcal{H})$:

1. If $\|\rho - \sigma\|_1 \leq \epsilon$, then $F(\rho, \sigma) \geq 1 - \epsilon$.
2. If $F(\rho, \sigma) \geq 1 - \epsilon$, then $\|\rho - \sigma\|_1 \leq 2\sqrt{\epsilon}$.

---

[7]Other ways to define a metric in terms of the fidelity are the following: One is the so-called *Bures distance*, which is defined as $D_B(\rho, \sigma) = \sqrt{2 - 2F(\rho, \sigma)}$, and the other is the *angle*, which is defined as $D_A(\rho, \sigma) = \arccos \sqrt{F(\rho, \sigma)}$.

It is also possible to define the fidelity in a different way using the concept of purification that was introduced in the previous section (see Definition 2.53). The fidelity can then be characterized in terms of the pure-state fidelity as given in (2.169) (for a proof see [3, Thm. 9.4]):

Theorem 2.61 (Uhlmann's Theorem) *Suppose $\rho_A, \sigma_A \in \mathcal{B}(\mathcal{H}_A)$ are states of a quantum system A and let R be a second quantum system which is a copy of A. Then*

$$F(\rho_A, \sigma_A) = \max_{|\psi_\rho\rangle, |\varphi_\sigma\rangle} |\langle \psi_\rho | \varphi_\sigma \rangle|^2, \tag{2.174}$$

where the maximization is over all purifications $|\psi_\rho\rangle$ of ρ_A and $|\varphi_\sigma\rangle$ of σ_A into $A \otimes R$.

Moreover, it can be shown (for example, in [2]) that for any fixed purification $|\psi_\rho\rangle$ of ρ, there is a purification $|\varphi_\sigma\rangle$ of σ that realizes the above maximum.

References

1. Gilchrist, A., Langford, N.K., Nielsen, M.A.: Distance measures to compare real and ideal quantum processes. Phys. Rev. A **71**(6) (2005). https://doi.org/10.1103/physreva.71.062310
2. Jozsa, R.: Fidelity for mixed quantum states. J. Mod. Opt. **41**(12), 2315–2323 (1994). https://doi.org/10.1080/09500349414552171
3. Nielsen, M.A., Chuang, I.L.: Quantum Computation and Quantum Information. Cambridge University Press, Cambridge (2000)
4. Werner, R.F.: Mathematical methods of quantum information theory (Lecture) (2017). https://www.youtube.com/playlist?list=PLDfPUNusx1EoBAn8vXYjcF95R7mI_eR6o
5. Wilde, M.M.: Quantum Information Theory, 2nd edn. Cambridge University Press, Cambridge (2017). https://doi.org/10.1017/9781316809976
6. Wootters, W.K., Zurek, W.H.: A single quantum cannot be cloned. Nature **299**(5886), 802–803 (1982). https://doi.org/10.1038/299802a0

Information and Entropies

3

Abstract

Entropies are an important tool to quantify information and uncertainty and, as such, a crucial part of security proofs of quantum key distribution protocols. In general, the quantum entropy quantifies the amount of uncertainty we have of the state of a quantum mechanical system. In the context of a quantum key distribution protocol, the entropy can be used, for instance, to estimate how much information an adversary has about the key that Alice and Bob try to establish. However, entropies do not only appear in the context of quantum information theory. They also appear in the study of classical information theory, and studying the classical counterparts helps developing an intuition for the language and the way of thinking we need when working with entropies. Therefore, in the first part of this chapter we concentrate on classical entropies and tasks in classical information theory. In the second part, we discuss quantum entropies and the similarities and differences to their classical counterparts. We also present an entropic version of uncertainty relations, which will later turn out to be a valuable tool for proving the security of quantum communication protocols.

3.1 Classical Entropies

For a moment, we will leave the quantum world behind and consider a purely classical world. More precisely, we want to study classical *entropies* and information theory. If you are interested in a more detailed treatment of this topic, great introductory books are [6] and [2]. It already became clear in the previous chapter that we are interested in *information*, usually encoded into bits. Since we are in a purely classical world for the moment, we only mean classical bits here. The issue of how to store and transmit classical information was first addressed by Claude

© The Author(s), under exclusive license to Springer Nature Switzerland AG 2021
R. Wolf, *Quantum Key Distribution*, Lecture Notes in Physics 988,
https://doi.org/10.1007/978-3-030-73991-1_3

Shannon in 1948 [12]. In his seminal work, he studied the two key questions of information theory:

(Q1) How many bits are required to reliably compress a given amount of infor-
       mation, that is, compress it in a way such that it can be later recovered with
       arbitrarily low probability of error?
(Q2) How much information can be reliably transmitted through a given communi-
       cation channel?

By answering these questions Shannon laid the foundation of classical information theory. However, to be able to do so, the first question that has to be addressed is an even more fundamental one: How do we quantify information?[1]

In the following, a lot of the arguments use notions and rules from probability theory, which are summarized in Appendix A. Consider a random variable X, where each realization x belongs to an alphabet $\mathcal{X}$. The probability that a specific realization $x \in \mathcal{X}$ occurs is denoted as $p_X(x)$. The information content of a specific realization x[2] is then given by the function

$$i(x) = -\log(p_X(x)), \tag{3.1}$$

where the logarithm is taken to base two (as usually in information theory), which indicates that we measure information in units of bits. The function (3.1) is plotted in Fig. 3.1. Why did we choose this function to quantify information? The reason is that it has several nice features that fit our needs very well:

1. The information content (or surprise) only depends on the probability of the event
    that a specific realization x occurs, and not on the label itself. For instance, a
    random variable that takes the values 0 and 1 with respective probabilities $\frac{1}{3}$ and
    $\frac{2}{3}$ should contain the same amount of information as a random variable that takes
    the values $+$ and $-$ with respective probabilities $\frac{1}{3}$ and $\frac{2}{3}$.
2. The function is continuous in the variable p.
3. As shown in Fig. 3.1, the function behaves exactly as we would expect from a
    measure of surprise: It is high for unlikely events and low for those with a higher
    probability. Also, the function is non-negative for every realization x.
4. The function is additive (which is due to the choice of the logarithm): Suppose
    that the information source produces two realizations x_1 and x_2 of the random
    variable X independently of each other. Intuitively, learning about both of these
    realizations at once should have the same amount of information as the summing

---

[1]There are two complementary ways of phrasing this question: We can ask how much uncertainty we have about a random variable X *before* learning about it, or, from a different point of view, we can ask how much information we have gained *after* learning about X.

[2]In other words, (3.1) quantifies how surprised we are to see a particular realization x to appear.

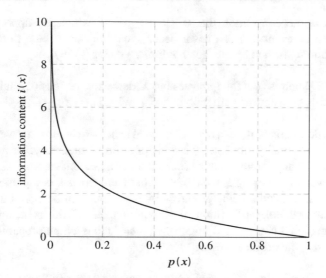

Fig. 3.1 Information content as a function of the probability p. Lower-probability events have a higher information content (or, stated differently, are more surprising), while more likely events have a lower information content

up the individual information contents of the two realizations:

$$i(x_1, x_2) = -\log\left(p_X(x_1, x_2)\right)$$
$$= -\log\left(p_X(x_1)p_X(x_2)\right)$$
$$= -\log\left(p_X(x_1)\right) - \log\left(p_X(x_2)\right)$$
$$= i(x_1) + i(x_2).$$

The joint probability distribution $p_X(x_1, x_2)$ factors as $p_X(x_1)p_X(x_2)$ because we assume that the two instances of the random variable X are independent of each other (which is the case if the information source is memoryless).

While the function in (3.1) is a suitable measure of information for a particular realization x of a random variable X, it does not capture the general amount of information that the random variable X possesses. To obtain a notion for the latter case, we consider the expectation value of the function $i(x)$ taken over all realizations x of X. This is referred to as the *entropy* or the *Shannon entropy* of the random variable X.

Definition 3.1 The entropy of a discrete random variable X with probability distribution $p_X(x)$ is defined as

$$H(X) = -\sum_x p_X(x) \log\left(p_X(x)\right). \tag{3.2}$$

Here, we use the convention that $0 \cdot \log(0) = 0$, which is supported by the intuition that an event that can never occur should not contribute to the entropy. It is furthermore justified by the fact that $\lim_{\epsilon \to 0} \epsilon \cdot \log(\epsilon) = 0$.

Exercise 3.2 Suppose the random variable X describes the toss of a fair coin, i.e., heads and tails both appear with probability $\frac{1}{2}$. What is $H(X)$?

With the definition of the Shannon entropy at hand, we can now answer one of the questions that originally motivated the study of information theory: To answer (Q1), consider an information source that is described by a random variable X. *Shannon's noiseless coding theorem* [12] then tells us that there exists a compression scheme such that the information being produced by the source can be stored using $H(X)$ bits per source symbol. Furthermore, the theorem states that this is the optimal case: if the information is compressed any further, then there is a high probability of error when retrieving the original message.[3]

Example 3.3 For a concrete example of data compression and Shannon's noiseless coding theorem, consider an information source X whose outputs are chosen randomly from a set of four symbols $\{A, B, C, D\}$ with respective probabilities $\frac{1}{2}, \frac{1}{4}, \frac{1}{8}, \frac{1}{8}$. Without any data compression, one needs 2 bits for each use of the source to encode the output symbol produced by the source:

symbol: A B C D
codeword: 00 01 10 11

In this case, the expected length of an individual codeword is two (since every codeword in this encoding scheme is of length two).

Is it possible to improve this scheme in a way that, on average, we need fewer bits per codeword? A common strategy is to take advantage of the skewed nature of the probability distribution: One encodes more likely symbols with short codewords, while encoding less likely symbols with longer codewords,[4] for example, the scheme

symbol: A B C D
codeword: 0 10 110 111

---

[3]The original theorem is a bit more technical, but we do not need it in all its details here to understand the idea.

[4]This strategy is, for example, employed in the famous Morse code, where the very common English letter "E" is encrypted with a single dot, while the less common letter "J" is encoded with a dot and three dashes.

gives an advantage over the previous scheme, which can be seen when we calculate the average length of a codeword:

$$\frac{1}{2} \cdot 1 + \frac{1}{4} \cdot 2 + \frac{1}{8} \cdot 3 + \frac{1}{8} \cdot 3 = \frac{7}{4}. \tag{3.3}$$

This value actually matches the entropy of the information source:

$$H(X) = -\frac{1}{2} \cdot \log\left(\frac{1}{2}\right) - \frac{1}{4} \cdot \log\left(\frac{1}{4}\right) - \frac{1}{8} \cdot \log\left(\frac{1}{8}\right) - \frac{1}{8} \cdot \log\left(\frac{1}{8}\right) = \frac{7}{4}. \tag{3.4}$$

Note that we could have employed a scheme that uses even fewer bits on average, e.g., $a \to 0$, $b \to 1$, $c \to 01$, $d \to 10$. However, this scheme has the disadvantage that a coded sequence like 00110100 is not uniquely decodable: It can, for example, be divided into codewords as $0\ 01\ 10\ 10\ 0$ or $0\ 0\ 1\ 10\ 1\ 0\ 0$ (and these are just two of the possibilities). On the other hand, the scheme described above provides only one way of dividing the sequence into individual codewords, namely $0\ 0\ 110\ 10\ 0$. This is an important property for the compressed data to be *reliably* decodable.

Example 3.4 A special case of the entropy occurs when we consider a random variable X that has a two-outcome set $\{0, 1\}$. The probabilities are then given by the distribution $p_X(0) = p$, $p_X(1) = 1 - p$ with $p \in [0, 1]$. The entropy in this case is called the *binary entropy*, usually denoted as h_2, which is a function of the parameter p (depicted in Fig. 3.2):

$$h_2(p) = -p \log(p) - (1 - p) \log(1 - p). \tag{3.5}$$

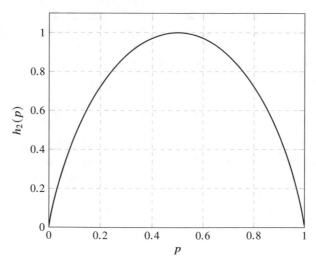

Fig. 3.2 Binary entropy. The binary entropy h_2 given in (3.5) as a function of the parameter p. The plot shows that it reaches its maximum at $p = 0.5$

Note that $h_2(p) = h_2(1 - p)$. The binary entropy quantifies, for example, the number of bits we learn from the outcome of a coin flip. If the coin is unbiased, i.e., $p = \frac{1}{2}$, then $h_2(p)$ reaches its maximum at one bit (which can be seen in the plot in Fig. 3.2). The other extreme is a deterministic coin ($p = 0$ or $p = 1$), where $h_2(p) = 0$, so we do not learn anything from the outcome of the coin flip.

Exercise 3.5 Consider an eight-sided die where every outcome i has the same probability, i.e., $p(i) = \frac{1}{8}$ for every $i \in \{1, \ldots, 8\}$.

1. What is the entropy of the die?
2. Suppose you want to send the outcome of rolling the die. What is the most efficient way to encode the outcomes? *Hint: Use that the entropy can be interpreted as average number of bits required to transmit the outcome of the die.*

3.1.1 Mathematical Properties

The entropy function defined in Definition 3.1 has several important properties:

1. The entropy function is non-negative for any random variable X with probability distribution $p_X(x)$:

$$H(X) \geq 0. \tag{3.6}$$

   Intuitively, this makes sense because the entropy represents the number of bits we learn upon learning a random variable X. Hence, the least amount of bits we can learn is zero since we can never learn a negative amount of bits. This intuition is supported mathematically by the fact that the entropy is the expected information content of $i(X)$, which itself is a non-negative quantity.
2. The entropy function is concave in the probability distribution $p_X(x)$, which means that

$$H(\lambda p_1 + (1 - \lambda)p_2) \geq \lambda H(p_1) + (1 - \lambda)H(p_2). \tag{3.7}$$

   Intuitively, this implies that the entropy increases under averaging. From the plot in Fig. 3.2 it is easy to see that the binary entropy function has this property.
3. The entropy is invariant under permutations of the realizations of the random variable X. This directly follows from the fact that the entropy only depends on the probabilities of the realizations, not on the values of the realizations themselves.

4. The entropy vanishes if and only if X is a deterministic variable. Since we do not have any uncertainty about a deterministic variable, we cannot gain any information by learning about it. It is also easy to show this property mathematically: If X is a deterministic random variable, the probability distribution is given by $p_X(x) = \delta_{x,x_0}$, where x_0 is the specific value that occurs with probability one. The entropy is then

$$H(X) = -\sum_x \delta_{x,x_0} \log(\delta_{x,x_0}) = -\log(1) = 0.$$

On the other hand, if $H(X) = 0$, this implies that for each $x \in X$, we have that $p_X(x) \log\big(p_X(x)\big) = 0$. This in turn implies that $p_X(x)$ is either equal to one or equal to zero for all $x \in X$. Because of the requirement that $\sum_x p_X(x) = 1$, the only possible choice is that $p_X(x) = 1$ for exactly one $x \in X$ and $p_X(x) = 0$ for all the other $x \in X$.

5. Given a random variable X that takes values in the alphabet X, the maximum value the entropy can take is given by

$$H(X) \leq \log |X|, \tag{3.8}$$

where $|X|$ is the cardinality of the alphabet. Equality holds if and only if X is a uniform random variable, i.e., a random variable with a uniform probability distribution.[5]

Exercise 3.6 Proof that the binary entropy $h_2(p)$ defined in (3.5) is a concave function, i.e., show that

$$h_2(\lambda p_1 + (1 - \lambda)p_2) \geq \lambda h_2(p_1)(1 + \lambda)h_2(p_2), \tag{3.9}$$

where $0 \leq \lambda, p_1, p_2 \leq 1$.

3.1.2 Conditional Entropy

Suppose that we have two random variables, X and Y. How can we relate the information content of X to the information content of Y? Consider the scenario depicted in Fig. 3.3: Alice possesses a random variable X that Bob wants to learn about. In the beginning, his uncertainty of X is modelled by the entropy $H(X)$. Alice then starts to send realizations x of X over a (possibly noisy) channel to Bob.

---

[5]For a proof of this statement we first have to introduce the relative entropy; therefore, it will be given later in this chapter.

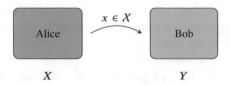

$$X \qquad\qquad\qquad\qquad Y$$

Fig. 3.3 Bob learns about Alice's random variable X. By sending over realizations x of the random variable X that Alice possesses, Bob learns about it in the form of side information, modelled by another random variable Y

Bob receives some information y that belongs to another random variable Y (since the output of a noisy channel cannot always be predicted with certainty), i.e., he has some "side information" about X. Using this information, he would now like to estimate his uncertainty about X. The entropy of the random variable X given a *particular* realization $y \in \mathcal{Y}$ as side information is

$$H(X|Y = y) = - \sum_x p_{X|Y}(x|y) \log\left(p_{X|Y}(x|y)\right), \tag{3.10}$$

where $p_{X|Y}(x|y)$ describes the probability that the event x occurs when we already know that the event y has occurred or will occur.

Before we can define the conditional entropy, we need to introduce another notion of probability, namely the *joint probability* $p_{X,Y}(x, y)$. It describes the probability that both events x and y occur, where x is a realization of the random variable X and y is a realization of the random variable Y. It is clear that if X and Y are independent of each other, the probability distribution just factors:

$$p_{X,Y}(x, y) = p_X(x) p_Y(y). \tag{3.11}$$

The joint probability is connected to the conditional probability in the following way:

$$p_{X|Y}(x|y) = \frac{p_{X,Y}(x, y)}{p_Y(y)} \tag{3.12}$$

for all $x \in \mathcal{X}$.

We can now answer the question we raised above: Suppose that Bob has access not only to one particular realization $y \in \mathcal{Y}$, but to the random variable Y. His uncertainty about Alice's random variable X is then given by the *conditional entropy*:

Definition 3.7 Let X and Y be discrete random variables with the joint probability distribution $p_{X,Y}(x, y)$. The conditional entropy $H(X|Y)$ is given by

$$H(X|Y) = \sum_y p_Y(y) H(X|Y = y) \tag{3.13}$$

$$= -\sum_y p_Y(y) \sum_x p_{X|Y}(x|y) \log\left(p_{X|Y}(x|y)\right) \tag{3.14}$$

$$= -\sum_{x,y} p_{X,Y} \log\left(p_{X|Y}(x|y)\right). \tag{3.15}$$

An application of the conditional entropy is presented in Example 3.8. Here, the entropy of a source that outputs a periodic string of bits is evaluated, conditioned on previously received bits.

Example 3.8 Suppose we have a source that outputs a periodic string of bits: ...01001001001.... We want to calculate the entropy of the source while taking into account the bits we have received previously. In other words, we want to estimate how much information we gain by receiving a bit from the source, conditioned on the number of bits we have already received. There are three cases that can appear:

1. **No known symbols.** When no previous symbol is known, the probability distribution of the symbols is given by the frequency of the respective symbol in the sequence:

$$p_X(x = 0) = \frac{2}{3}, \quad p_X(x = 1) = \frac{1}{3}.$$

Using the definition of the Shannon entropy, we find that

$$H(X) = -\frac{2}{3} \log\left(\frac{2}{3}\right) - \frac{1}{3} \log\left(\frac{1}{3}\right) \approx 0.918 \text{ bit.}$$

2. **One known symbol.** In this case we have some side information Y (given by the known symbol); therefore, we need to calculate the conditional entropy

$$H(X|Y) = \sum_y p_Y(y) H(X|Y = y).$$

The probability distribution $p_Y(y)$ is simply the probability that the respective symbol appears in the sequence, i.e., it is given by the one we listed above. To calculate the conditional entropy $H(X|Y = y)$, we need to know the probability that a symbol appears given that we know the value of the previous symbol, which is the conditional probability distribution:

$p_{X|Y}(x|y)$	$x = 0$	$x = 1$
$y = 0$	1/2	1/2
$y = 1$	1	0

If the previous symbol is one, we know that the next symbol must be zero since there are never two ones in a row in the sequence given above. We can then calculate the conditional entropy:

$$H(X|Y) = -\frac{2}{3} \cdot H(X|Y = 0) - \frac{1}{3} \cdot H(X|Y = 1)$$

$$= \frac{2}{3} \cdot 1 + \frac{1}{3} \cdot 0 = \frac{2}{3} \text{ bit.}$$

As expected, after having received some information from the source the entropy is decreased. In other words, receiving a bit of information has decreased our uncertainty of the source X.

3. **Two known symbols.** If we already know two symbols, we should be certain about which symbol comes next. This is indeed reflected in the conditional entropy. The conditional probability distribution is now given by

$p_{X|Y}(x|y)$	$x = 0$	$x = 1$
$y = 00$	0	1
$y = 01$	1	0
$y = 10$	1	0
$y = 11$	–	–

There can never be two ones in a row; therefore, there are no probabilities for the last row. The table includes now only ones and zeros; therefore, there is no uncertainty left and $H(X|Y = y) = 0$ for all y.

There is an obvious relation between the conditional entropy $H(X|Y)$ and the margin entropy $H(X)$: Suppose $H(X)$ is the uncertainty that Bob has about Alice's random variable X before he receives any information and $H(X|Y)$ is the uncertainty he has after getting some side information Y. Intuitively, his uncertainty after getting some side information should not increase. At worst, the side information does not have any effect on the uncertainty. This is formalized by the following theorem:

Theorem 3.9 *Conditioning does not increase the entropy of a random variable: Consider two random variables X and Y. Then*

$$H(X) \geq H(X|Y) \tag{3.16}$$

with equality if and only if X and Y are independent random variables.

Proof First, note that the following equality of probabilities holds (see Appendix A):

$$\sum_{y \in \mathcal{Y}} p_{X,Y}(x, y) = p_X(x). \tag{3.17}$$

We can then prove the theorem by the following sequence of equalities:

$$H(X) - H(X|Y) = -\sum_{x} p_X \log\left(p_X(x)\right) + \sum_{x,y} p_{X,Y} \log\left(p_{X|Y}(x|y)\right) \tag{3.18}$$

$$= -\sum_{x,y} p_{X,Y}(x, y) \log\left(p_X(x)\right) - p_{X,Y}(x, y) \log\left(p_{X|Y}(x|y)\right) \tag{3.19}$$

$$= \sum_{x,y} p_{X,Y}(x, y) \log\left(\frac{p_{X,Y}(x, y)}{p_X(x)p_Y(y)}\right) \tag{3.20}$$

$$= \sum_{x,y} p_X(x)p_Y(y) \underbrace{\frac{p_{X,Y}(x, y)}{p_X(x)p_Y(y)}}_{\equiv z} \log\left(\underbrace{\frac{p_{X,Y}(x, y)}{p_X(x)p_Y(y)}}_{\equiv z}\right). \tag{3.21}$$

The function $\varphi(z) = z \log(z)$ is a convex function; hence, we can use Jensen's inequality. This states that for a convex function $f(x)$ and a discrete random variable X, it holds that $f(E[X]) \leq E[f(X)]$. Applied to the above expression, this yields

$$H(X) - H(X|Y) = \sum_{x,y} p_X(x)p_Y(y)\, \varphi\left(\frac{p_{X,Y}(x, y)}{p_X(x)p_Y(y)}\right) \tag{3.22}$$

$$\geq \varphi\left(\sum_{x,y} p_X(x)p_Y(y) \frac{p_{X,Y}(x, y)}{p_X(x)p_Y(y)}\right) \tag{3.23}$$

$$= \varphi(1) \tag{3.24}$$

$$= 0. \tag{3.25}$$

Note that the only inequality appears in the second line. When does this become an equality, i.e., when is $H(X) - H(X|Y) = 0$? If X and Y are statistically independent, $p_{X,Y}(x, y) = p_X(x)p_Y(y)$ and the argument of φ in (3.22) directly becomes 1. On the other hand, if $H(X) = H(Y|X)$, then (3.20) has to be zero. The log function is only zero if its argument equals one; hence, $p_{X,Y}(x, y) = p_X(x)p_Y(y)$ and therefore, X and Y are statistically independent. $\square$

3.1.3 Joint Entropy

The definition of entropy can be extended to the case where we have two random variables X and Y and have no knowledge about either of them. The quantity that describes our uncertainty about the pair (X, Y) is the *joint entropy*:

Definition 3.10 Let X and Y be discrete random variables with the joint probability distribution $p_{X,Y}(x, y)$. The joint entropy $H(X, Y)$ is then given by

$$H(X, Y) = - \sum_{x,y} p_{X,Y}(x, y) \log \left(p_{X,Y}(x, y) \right). \tag{3.26}$$

We can now establish several useful inequalities regarding the relation between the joint entropy $H(X, Y)$, the conditional entropies $H(X|Y)$ and $H(Y|X)$, and the margin entropies $H(X)$ and $H(Y)$:

Theorem 3.11 *Given two discrete random variables X and Y with joint probability distribution $p_{X,Y}(x, y)$, the following relations hold:*

$$H(X, Y) = H(X) + H(Y|X) = H(Y) + H(X|Y). \tag{3.27}$$

Proof We can easily prove these relations by considering the multiplicative relation between probabilities $p_{X,Y}(x, y) = p_{Y|X}(y|x)p_X(x)$:

$$H(X, Y) = - \sum_{x,y} p_{X,Y}(x, y) \log \left(p_{X,Y}(x, y) \right)$$

$$= - \sum_{x,y} p_{X,Y}(x, y) \log \left(p_{Y|X}(y|x)p_X(x) \right)$$

$$= - \sum_{x} p_X(x) \log \left(p_X(x) \right) - \sum_{x,y} p_{X,Y}(x, y) \log \left(p_{Y|X}(y|x) \right)$$

$$= H(X) + H(Y|X).$$

An analogous calculation shows the second relation. $\square$

The result of the above theorem allows us to formulate some practical chain rules for entropies:

Corollary 3.12 (Chain Rule for Conditional Entropies) *For discrete random variables* $X_1, X_2, \ldots, X_n$ *and* Y, *the following chain rule for the conditional entropy holds:*

$$H(X_1, X_2, \ldots, X_n|Y) = \sum_{i=1}^{n} H(X_i|Y, X_1, \ldots, X_{i-1}). \tag{3.28}$$

Proof We will prove this statement by induction over n. For $n = 2$ we can simply follow from the definitions that

$$H(X_1, X_2|Y) = H(X_1, X_2, Y) - H(Y) \tag{3.29}$$

$$= H(X_1, X_2, Y) - H(X_1, Y) + H(X_1, Y) - H(Y) \tag{3.30}$$

$$= H(X_2|Y, X_1) + H(X_1, Y). \tag{3.31}$$

Now assume that the result holds for general n, and show that it holds for $n + 1$. We can use the result from the calculation above for $n = 2$ to write

$$H(X_1, \ldots, X_{n+1}|Y) = H(X_2, \ldots, X_{n+1}|Y, X_1) + H(X_1|Y). \tag{3.32}$$

Apply the induction hypothesis, namely that the result holds for general n, to the right hand side:

$$H(X_1, \ldots, X_{n+1}|Y) = \sum_{i=2}^{n+1} H(X_i|Y, X_1, \ldots, X_{i-1}) + H(X_1|Y) \tag{3.33}$$

$$= \sum_{i=1}^{n+1} H(X_i|Y, X_1, \ldots, X_{i-1}), \tag{3.34}$$

so the statement also holds for $n + 1$. □

Corollary 3.13 (Chain Rule for Joint Entropies) *For discrete random variables* $X_1, X_2, \ldots, X_n$, *the following chain rule for the joint entropy holds*

$$H(X_1, X_2, \ldots, X_n) = H(X_1) + H(X_2|X_1) + \cdots + H(X_n|X_{n-1}, \ldots, X_1). \tag{3.35}$$

Proof Combining the result of Theorem 3.11 with the result of Corollary 3.12 yields

$$H(X_1, \ldots, X_n) = H(X_1) + H(X_2, \ldots, X_n | X_1) \tag{3.36}$$

$$= H(X_1) + \sum_{i=2}^{n} H(X_i | X_1, X_2, \ldots, X_{i-1}), \tag{3.37}$$

which is the statement of the corollary.   □

Note that Theorem 3.11 implies that $H(X, Y) \geq H(X)$ and $H(X, Y) \geq H(Y)$ since the conditional entropy is always non-negative (this will be different in the quantum case!). Theorem 3.11 also implies that $H(X, Y) = H(X) + H(Y|X) \leq H(X) + H(Y)$, where the inequality follows directly from Theorem 3.9, i.e., the fact that conditioning cannot increase the entropy. It can be generalized to an arbitrary number of random variables:

Theorem 3.14 *For discrete random variables* $X_1, \ldots, X_n$, *the joint entropy is subadditive, which means that*

$$H(X_1, \ldots, X_n) \leq \sum_{i=1}^{n} H(X_i). \tag{3.38}$$

Equality holds if and only if the $X_1, \ldots, X_n$ *are independent random variables.*

Proof This statement is a direct application of the chain rule for joint entropies (Corollary 3.13) and the fact that conditioning does not increase the entropy (Theorem 3.9):

$$H(X_1, \ldots, X_n) = H(X_1) + \sum_{i=1}^{n} \underbrace{H(X_i | X_1, \ldots, X_{i-1})}_{\leq H(X_i)} \tag{3.39}$$

$$\leq \sum_{i=1}^{n} H(X_i). \tag{3.40}$$

Equality is achieved if and only if $H(X_i | X_1, \ldots, X_{i-1}) = H(X_i)$ for all i, which, according to Theorem 3.9, holds if and only if X_i are independent random variables.
                                                                                □

3.1.4 Mutual Information

With the entropies we have defined above, we can now define an entropic measure that quantifies the *mutual information* that two parties posses. Consider a scenario where Alice holds a random variable X and Bob has a random variable Y.

Definition 3.15 For two discrete random variables X and Y with joint probability distribution $p_{X,Y}(x, y)$, the mutual information $I(X : Y)$ is defined as

$$I(X : Y) = H(X) - H(X|Y). \tag{3.41}$$

The interpretation of the formula above is the following: $H(X)$ represents the uncertainty we have about the random variable X. Since $H(X|Y)$ tells us how much uncertainty there is left about X after we have learned about Y, their difference gives the information that X and Y share. An equivalent formula for the mutual information is

$$I(X : Y) = H(Y) - H(Y|X). \tag{3.42}$$

The definition of the mutual information directly implies that if the two random variables X and Y are statistically independent of each other, they do not have any mutual information, i.e., learning about Y does not reduce the uncertainty we have about X (and vice versa). This directly follows from the fact that, for two independent random variables X and Y, $H(X|Y) = H(X)$ and therefore $I(X : Y) = 0$. This, along with the non-negativity of the mutual information (which is a direct consequence of Theorem 3.9), is summed up by the following theorem:

Theorem 3.16 *For two random variables X and Y, the mutual information is non-negative:*

$$I(X : Y) \geq 0 \tag{3.43}$$

with equality if and only if X and Y are independent random variables.

Exercise 3.17 Consider a random variable X with alphabet $\mathcal{X}$. For a subset $\mathcal{S} \subseteq \mathcal{X}$, let Y be the random variable that represents the answer to the question whether or not X lies in $\mathcal{S}$, i.e.,

$$Y = \begin{cases} 1 & \text{if } X \in \mathcal{S} \\ 0 & \text{if } X \notin \mathcal{S}. \end{cases} \tag{3.44}$$

Calculate the decrease in uncertainty about X, which is $I(X : Y) = H(X) - H(X|Y)$, in terms of the probability $\alpha = P[X \in \mathcal{S}]$.

We can also consider the case where we want to know the mutual information of two random variables X and Y, given that we have some side information in the form of another random variable Z. This is called the *conditional mutual information* and is given by

$$I(X : Y|Z) = H(Y|Z) - H(Y|X, Z) \tag{3.45}$$

$$= H(X|Z) - H(X|Y, Z) \tag{3.46}$$

$$= H(X|Z) + H(Y|Z) - H(X, Y|Z). \tag{3.47}$$

Exercise 3.18 Show that the conditional mutual information is non-negative:

$$I(X : Y|Z) \geq 0. \tag{3.48}$$

Exercise 3.19 Show the following chain rule for mutual information:

$$I(X_1, X_2, \ldots, X_n : Y) = \sum_{i=1}^{n} I(X_i : Y|X_{i-1}, \ldots, X_1), \tag{3.49}$$

where for $i = 1$ the term in the sum is defined to be $I(X_1 : Y)$.

In Fig. 3.4, the relationship between the entropies that we have introduced so far is depicted: The lighter, green circle represents the marginal entropy $H(X)$ and therefore represents the uncertainty we have about a random variable X. The darker, blue circle represents the entropy of another random variable Y. Their intersection is the mutual information of these two random variables (note that, if X and Y are independent, the two circles would be disjoint sets). The green circle without the

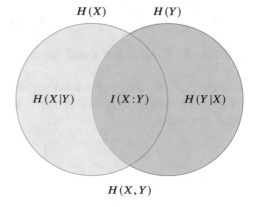

Fig. 3.4 Relationship between different entropies. The green (left) and blue (right) circles represent the entropy of the random variables X and Y, respectively. Their overlap is the mutual information of X and Y, and the green (blue) circle without the overlap is the entropy of X (Y) conditioned on Y (X). Both circles together represent the joint entropy of X and Y

overlap represents the conditional entropy $H(X|Y)$, which is the uncertainty that is left after we have learned about the random variable Y (analogously for the blue circle without the overlap). The area that includes both circles is the joint entropy $H(X, Y)$.

We can now give the answer to the second of Shannon's questions (Q2), which asked how much information can be reliably transmitted through a communication channel, which is covered by *Shannon's noisy coding theorem* (also called *Shannon's channel coding theorem*). Consider the following scenario, which is similar to the one we have seen in Fig. 3.3: Alice possesses a random variable X. She sends information about the random variable (i.e., realizations of X) over to Bob, using a noisy classical communication channel $\mathcal{N}$. The information that Bob has, i.e., the output of the noisy channel $\mathcal{N}$, is represented by a random variable, denoted as Y, because it is not possible to always predict the output of the channel with certainty. The number of bits that can be reliably transmitted per use of the channel is then given by the *capacity* $C(\mathcal{N})$ of the channel $\mathcal{N}$:

$$C(\mathcal{N}) = \max_{p_X(x)} I(X : Y),$$

where $p_X(x)$ is the probability distribution according to which Alice chooses realizations of her random variable X. Depending on which probability distribution she chooses, the mutual information between X and Y varies. However, the capacity of the channel should not depend on Alice's choice of $p_X(x)$; therefore, we take the maximum over all possible probability distributions.[6]

Exercise 3.20 The binary erasure channel is a classical channel where the input bit is erased with a certain probability p and transmitted correctly with probability $1 - p$, as depicted below. Calculate the capacity of this channel.

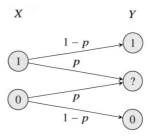

---

[6]Note that, when the probability distribution $p_{X|Y}(x|y)$ is fixed, $I(X : Y)$ is concave in $p_X(x)$; hence, there always exists a distribution $p_X^*(x)$ that maximizes $I(X : Y)$.

3.1.5 Relative Entropy

The last classical entropy we want to introduce here is the relative entropy, also known as *Kullback–Leibler divergence*. It quantifies how far one probability distribution $p(x)$ is from another probability distribution $q(x)$.

Definition 3.21 Suppose X is a finite set and $p(x)$ and $q(x)$ are two probability distributions over the set X. The relative entropy of $p(x)$ to $q(x)$ is

$$D\big(p(x)\|q(x)\big) = \sum_x p(x) \log\left(\frac{p(x)}{q(x)}\right) = -H(X) - \sum_x p(x) \log(q(x)).$$
(3.50)

We define $-p(x)\log(0) = +\infty$ if $p(x) > 0$.

Although the relative entropy is not a distance measure in the strict mathematical sense (because it is not symmetric in its arguments), the following theorem gives some motivation for why it can still be useful in order to estimate the difference between two probability densities:

Theorem 3.22 *Given two probability distributions $p(x)$ and $q(x)$ over some alphabet X, the relative entropy of $p(x)$ to $q(x)$ is non-negative:*

$$D\big(p(x)\|q(x)\big) \geq 0.$$
(3.51)

Equality holds if and only if $p(x) = q(x)$ for all $x \in X$.

Proof Note that $\log x \ln 2 = \ln x \leq x - 1$ for all positive x with equality if and only if $x = 1$.[7] We do not prove this formula here, but the plot in Fig. 3.5 suggests that it holds for all positive x, and also indicates that they are equal if and only if $x = 1$. Reformulating this formula slightly yields

$$- \log x \geq \frac{1 - x}{\ln 2}.$$
(3.52)

We can then consider the following relations:

$$D\big(p(x)\|q(x)\big) = - \sum_x p(x) \log\left(\frac{q(x)}{p(x)}\right)$$

$$\geq \frac{1}{\ln 2} \sum_x p(x) \left(1 - \frac{q(x)}{p(x)}\right)$$

---

[7]Here, ln denotes the natural logarithm, i.e., the logarithm to base e.

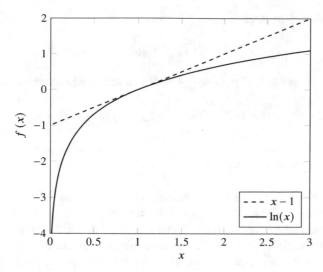

Fig. 3.5 Comparing the functions $\ln(x)$ and $x - 1$. The plot shows that $\ln(x) \leq x - 1$ for all positive x, and equality holds if and only if $x = 1$

$$= \frac{1}{\ln 2} \left(\sum_x p(x) - \sum_x q(x) \right)$$

$$= 0.$$

Equality only depends on the operation in the second line. As we have seen above, equality here holds if and only if $\frac{q(x)}{p(x)} = 1$ for all x, which implies that the two distributions are identical. $\quad\square$

Using Theorem 3.22, we can now give a proof of the fifth mathematical property of the entropy, which states that for a random variable X,

$$H(X) \leq \log |\mathcal{X}|, \tag{3.53}$$

where $|\mathcal{X}|$ is the cardinality of the alphabet and equality holds if and only if X is a uniform random variable. The proof works as follows: Suppose $p_X(x)$ is the probability distribution for the random variable X over $d \equiv |\mathcal{X}|$ outcomes and $q(x) = \frac{1}{d}$ denotes a uniform distribution over these outcomes. Then

$$D\big(p(x)\|\tfrac{1}{d}\big) = -H(X) - \sum_x p(x) \log \left(\frac{1}{d} \right) = \log(d) - H(X). \tag{3.54}$$

Non-negativity of the relative entropy implies

$$H(X) \leq \log(d), \tag{3.55}$$

with equality if and only if $p(x) = \frac{1}{d}$, i.e., X is a uniform random variable.

Exercise 3.23 Show that $I(X:Y) = D\big(p_{X,Y}(x,y) \| p_X(x)p_Y(y)\big)$.

3.2   Quantum Entropies

Classical entropies provide a tool to estimate the amount of information of a classical random variable and, moreover, to determine how much information two (or more) random variables share. Since our goal in these notes is to analyse quantum information, we would like to have similar tools to estimate the amount of information that is present in a quantum system. The mathematical quantity we are looking for is the *von Neumann entropy* (which we also simply call quantum entropy), which is the generalization of the Shannon entropy to quantum systems. The unit of information that we now use to determine the amount of quantum information in a system is that of *quantum bits*.

Recall that the classical entropy is a direct function of the probability distribution of the random variable, since it captures the classical uncertainty we have about a system. In the quantum world, classical uncertainty about a system as well as quantum uncertainty (that arises from the uncertainty principle) is captured by the notion of a density operator. Therefore, we expect a quantum analogue of the Shannon entropy to be a direct function of the density operator:

Definition 3.24 Let A be a quantum system that is prepared in a state $\rho_A \in \mathcal{B}(\mathcal{H}_A)$. The von Neumann entropy $H(A)_\rho$ of the state ρ_A is then defined as follows:

$$H(A)_\rho = -\mathrm{Tr}\,(\rho_A \log(\rho_A))\,. \tag{3.56}$$

We use the same notation H for the quantum entropy of a state as for the Shannon entropy of a classical random variable. Usually, it should be clear from the context, to which kind of entropy we are referring. We sometimes also use the notation $H(A)_\rho = H(\rho_A)$.

The quantum entropy has a useful relation to the eigenvalues of the density operator: Suppose we know the spectral decomposition of the state ρ_A, namely

$$\rho_A = \sum_j \lambda_j |\psi_j\rangle_A \langle\psi_j|, \tag{3.57}$$

where λ_j are the eigenvalues of ρ_A and $\{|\psi_j\rangle_A\}$ is the set of the corresponding pairwise orthogonal eigenvectors. The entropy of ρ_A is then

$$H(A)_\rho = -\sum_j \lambda_j \log(\lambda_j). \qquad (3.58)$$

Exercise 3.25 Show that the entropy $H(A)_\rho$ of the state ρ_A with spectral decomposition (3.57) is given by (3.58).

Exercise 3.26 Calculate the entropy of the state that corresponds to the ensemble $\{\{\frac{1}{4}, |0\rangle\}, \{\frac{1}{4}, |1\rangle\}, \{\frac{1}{4}, |+\rangle\}, \{\frac{1}{4}, |-\rangle\}\}$. Compare it to the classical entropy of the uniform distribution $\frac{1}{4}$.

As in the classical case, we define $0 \cdot \log(0) = 0$ and logarithms are usually taken to base two. The formula in (3.58) is the one that is usually used in calculations if we know the eigenvalues of the density matrix. Since the eigenvalues λ_j can be interpreted as a probability distribution (they fulfil the conditions $\sum_j \lambda_j = 1$ and $\lambda_j \geq 0$ for all j), this directly shows that the quantum entropy of a quantum state ρ is the same as the Shannon entropy for the probability distribution given by the eigenvalues λ_j.

The interpretation of quantum entropy is qualitatively very similar to the interpretation we gave for the classical entropy: Suppose that Alice prepares a quantum state $|\psi_x\rangle$ in her lab according to a probability distribution $p_X(x)$ that corresponds to a random variable X. Before she sends it to Bob, he does not know which state Alice will send, and therefore, from Bob's point of view, the expected density operator is

$$\rho = \sum_x p_X(x)|\psi_x\rangle\langle\psi_x|. \qquad (3.59)$$

The entropy of the state $H(A)_\rho$ then quantifies Bob's uncertainty about the state that Alice will send, or, from a different point of view, the amount of information that Bob gains when learning about the state.

3.2.1 Mathematical Properties of Quantum Entropy

The quantum entropy has several interesting mathematical properties that we discuss below. Some of the properties directly follow from the analogous properties in the classical case and the dependence of the entropy on the eigenvalues of the density operator:

1. The quantum entropy $H(\rho)$ is non-negative for any density operator ρ:

$$H(\rho) \geq 0. \tag{3.60}$$

   This follows directly from the non-negativity of the Shannon entropy.
2. The entropy vanishes if and only if the density operator is a pure state. The proof works similar to the one we gave to show that the Shannon entropy vanishes if and only if X is a deterministic variable:
   $\Rightarrow$ We can argue that if the entropy vanishes, then $\lambda_j \log(\lambda_j) = 0$ for all j, which is fulfilled either if $\lambda_j = 0$ or $\lambda_j = 1$. Since $\sum_j \lambda_j = 1$, there is exactly one j^* for which $\lambda_{j*} = 1$, so we have a pure state $\rho = |\psi_{j*}\rangle\langle\psi_{j*}|$.
   $\Leftarrow$ On the other hand, if the system is in a pure state $\rho = |\psi\rangle\langle\psi|$, then $H(\rho) = -\log(1) = 0$.
3. The maximum value of the quantum entropy of a state ρ is

$$H(\rho) \leq \log(d), \tag{3.61}$$

   where d is the dimension of the quantum system. This value is attained if ρ is the maximally mixed state. The proof is the same as the one we gave at the end of the previous section for the classical case (note that the eigenvalues of the maximally mixed state form a uniform probability distribution).
4. Consider a density operator $\rho_x \in \mathcal{B}(\mathcal{H})$ and a probability distribution $p_X(x)$. The quantum entropy is concave in the density operator:

$$H\left(\sum_x p_X(x)\rho_x\right) \geq \sum_x p_X(x)H(\rho_x). \tag{3.62}$$

   Here, the physical intuition is the same as in the classical case: The entropy (i.e., the uncertainty about a state) cannot decrease under mixing.
5. The entropy of a quantum state $\rho \in \mathcal{B}(\mathcal{H})$ is invariant under the action of an isometry $V : \mathcal{H} \to \mathcal{H}'$, i.e.,

$$H(\rho) = H(V\rho V^\dagger). \tag{3.63}$$

   This can be seen by doing a simple calculation: Suppose ρ has spectral decomposition $\rho = \sum_j \lambda_j |\psi_j\rangle\langle\psi_j|$. Then

$$V\rho V^\dagger = \sum_j \lambda_j V|\psi_j\rangle\langle\psi_j|V^\dagger \tag{3.64}$$

$$= \sum_j \lambda_j |\psi_j'\rangle\langle\psi_j'|, \tag{3.65}$$

where $\{|\psi'_j\rangle\}$ is an orthonormal basis such that $|\psi'_j\rangle = V|\psi_j\rangle$. Since the quantum entropy is only a function of the eigenvalues λ_j, it does not change when applying an isometry V.

In the following, we discuss quantum generalizations of the variants of the Shannon entropy that we have introduced in the previous section, such as conditional entropy, joint entropy, mutual information, and relative entropy. Although the definitions are quantum analogues of the classical quantities, we will see some surprising properties that clearly distinguish the quantum entropies from their classical counterparts.

3.2.2 Joint Quantum Entropy

We follow a slightly different order here than we did in the classical case. While in the previous section we first introduced the conditional entropy, here we have to begin with the joint quantum entropy since the definition of the conditional quantum entropy relies on it.

To formulate a quantum analogue of the classical joint entropy, we consider a composite system of the form $\mathcal{H}_A \otimes \mathcal{H}_B$. For a bipartite quantum state $\rho_{AB} \in \mathcal{B}(\mathcal{H}_A \otimes \mathcal{H}_B)$, we can then define the notion of the joint quantum entropy $H(AB)_\rho$ as a natural extension of the definition of quantum entropy:

Definition 3.27 Let $\rho_{AB} \in \mathcal{B}(\mathcal{H}_A \otimes \mathcal{H}_B)$ be the density operator for a bipartite system. The joint entropy $H(AB)_\rho$ is defined as

$$H(AB)_\rho = -\mathrm{Tr}\big(\rho_{AB} \log(\rho_{AB})\big). \tag{3.66}$$

This definition can analogously be defined for a composite system of three or more systems. Furthermore, there is also a notion of the joint entropy of a bipartite system $H(AB)_\rho$ if we have a tripartite (or multipartite) state $\rho_{ABC} \in \mathcal{B}(\mathcal{H}_A \otimes \mathcal{H}_B \otimes \mathcal{H}_C)$. The formula is the same as above, with the state being $\rho_{AB} = \mathrm{Tr}_C(\rho_{ABC})$. We will use this convention throughout these notes.

We will now encounter the first radical difference between classical and quantum entropies by studying the properties of the joint quantum entropy. Recall that in the classical case, the joint entropy is always greater or equal to the marginal entropies:

$$H(X, Y) \geq H(X), \qquad H(X, Y) \geq H(Y), \tag{3.67}$$

which follows from Theorem 3.11. In the quantum world, however, these inequalities do not always hold. In fact, for any pure bipartite state with Schmidt rank greater than one, the joint entropy vanishes, while the marginal entropies are equal, which represents one of the most fundamental differences between classical and quantum information:

Theorem 3.28 *Let* $|\psi\rangle_{AB} \in \mathcal{H}_A \otimes \mathcal{H}_B$ *be a pure bipartite state. Then the marginal entropies are equal:*[8]

$$H(A)_\psi = H(B)_\psi \qquad (3.68)$$

and the joint entropy vanishes

$$H(AB)_\psi = 0. \qquad (3.69)$$

Proof The most important ingredient in the proof is the Schmidt decomposition. Recall from Theorem 2.44 and the discussion afterwards that a pure bipartite state admits a representation of the form

$$|\psi\rangle_{AB} = \sum_i \lambda_i |e_i\rangle_A |f_i\rangle_B, \qquad (3.70)$$

with $\lambda_i > 0$ for all i and $\sum_i \lambda_i = 1$ and $\{|e_i\rangle_A\}$ is an orthonormal set of vectors on system A and $\{|f_i\rangle_B\}$ is an orthonormal set of vectors on B. The marginal states are then given by

$$\rho_A = \sum_i \lambda_i^2 |e_i\rangle_A \langle e_i|, \qquad \rho_B = \sum_i \lambda_i^2 |f_i\rangle_B \langle f_i|. \qquad (3.71)$$

Hence, both marginal states have the same eigenvalue, namely λ_i^2. Since the entropy is only a function of the eigenvalues, it follows that $H(A)_\psi = H(B)_\psi$. The joint entropy of the state $|\psi\rangle_{AB}$ vanishes by the same argument that we gave for the von Neumann entropy (see Item 2 of the mathematical properties). □

In the classical case, the joint entropy of two independent random variables is simply the sum of the margin entropies: $H(X, Y) = H(X) + H(Y)$ (see Theorem 3.14). The quantum analogue of this is the entropy of a product state $\rho_A \otimes \sigma_B$, where $\rho_A \in \mathcal{B}(\mathcal{H}_A)$ and $\sigma_B \in \mathcal{B}(\mathcal{H}_B)$. The quantum entropy is additive in this case:

$$H(\rho_A \otimes \sigma_B) = H(\rho_A) + H(\sigma_B). \qquad (3.72)$$

Exercise 3.29 Prove the formula in (3.72).

---

[8]Recall that this notation means that we calculate the entropy of the states $\rho_A = \mathrm{Tr}_B (|\psi\rangle_{AB} \langle\psi|)$ and $\rho_B = \mathrm{Tr}_A (|\psi\rangle_{AB} \langle\psi|)$, respectively.

Another interesting quantity arises when we consider states in quantum–classical hybrid systems. Recall from Eq. (2.141) that the density operator of a quantum–classical system $A \otimes Z$ is

$$\rho_{AZ} = \sum_z p_Z(z) \rho_A^z \otimes |z\rangle_Z \langle z|. \tag{3.73}$$

The joint entropy of this state then takes a specific form (the proof of this theorem can be found in [17, Thm. 11.2.2]):

Theorem 3.30 *The joint entropy $H(AZ)_\rho$ of a classical–quantum state ρ_{AZ} as given in (3.73) takes the form*

$$H(AZ)_\rho = H(Z) + \sum_z p_Z(z) H(\rho_A^z). \tag{3.74}$$

3.2.3 Conditional Quantum Entropy

After having defined the joint quantum entropy, we can use it to give a definition that generalizes the conditional entropy to quantum states. The formula for the conditional quantum entropy is inspired by the relation between classical conditional and joint entropy we have seen in Theorem 3.11:

Definition 3.31 Let $\rho_{AB} \in \mathcal{B}(\mathcal{H}_A \otimes \mathcal{H}_B)$ be a bipartite quantum state. The conditional quantum entropy $H(A|B)_\rho$ of ρ_{AB} is defined as

$$H(A|B)_\rho = H(AB)_\rho - H(B)_\rho. \tag{3.75}$$

As in the classical case, conditioning does not increase the quantum entropy, even if the conditioning system is quantum:

Theorem 3.32 *For a bipartite quantum state ρ_{AB}, the following holds for the marginal entropy $H(A)_\rho$ and the conditional quantum entropy $H(A|B)_\rho$:*

$$H(A)_\rho \geq H(A|B)_\rho. \tag{3.76}$$

Using the result of Theorem 3.30, we can directly give an expression for the conditional entropy $H(A|Z)_\rho$ of a classical–quantum state ρ_{AZ} if the conditioning is on the classical system Z: for a classical–quantum state ρ_{AZ} as given in (3.73),

$$H(A|Z)_\rho = H(AZ)_\rho - H(Z) \tag{3.77}$$

$$= H(Z) + \sum_z p_Z(z) H(\rho_A^z) - H(Z) \tag{3.78}$$

$$= \sum_z p_Z(z) H(\rho_A^z). \tag{3.79}$$

This formula is analogous to the classical formula given in (3.13).

We are now ready to encounter a fascinating phenomenon that clearly distinguishes quantum information theory from its classical counterpart that is even less intuitive than the one we studied in Theorem 3.28. We are talking about the fact that the conditional entropy can be negative! Recall the bipartite entangled state

$$|\Phi^+\rangle_{AB} = \frac{1}{\sqrt{2}} \left(|0\rangle_A |0\rangle_B + |1\rangle_A |1\rangle_B \right), \tag{3.80}$$

which is one of the Bell states that we introduced in the previous section. To compute the conditional entropy $H(A|B)_{\Phi^+}$ we need to calculate the joint entropy $H(AB)_{\Phi^+}$ as well as the marginal entropy $H(B)_{\Phi^+}$. Since $|\Phi^+\rangle_{AB}$ is a pure state, we know from Theorem 3.28 that the joint entropy vanishes

$$H(AB)_{\Phi^+} = 0. \tag{3.81}$$

To evaluate the marginal entropy $H(B)_{\Phi^+}$, we have to calculate the entropy of the marginal state on system B, which is the maximally mixed state π_B. From Item 3 of the mathematical properties of the quantum entropy, we know that the entropy of the maximally mixed state is given by

$$H(B)_{\Phi^+} = \log(d). \tag{3.82}$$

In our case, which is the qubit case, the dimension of the system is two; hence, $H(B)_{\Phi^+} = \log(2) = 1$. Combining these two results yields the conditional entropy:

$$H(A|B)_{\Phi^+} = H(AB)_{\Phi^+} - H(B)_{\Phi^+} = -1. \tag{3.83}$$

What do we make of such a result? What does it mean that the conditional entropy of a quantum state is negative? We can interpret it in the following way: Although we know precisely in which state the composite system is (since it is described by a pure state), it is possible to have less knowledge about the individual parts, which results in a negative conditional quantum entropy.[9] This is clearly a property of the quantum world that is impossible in the classical world.

We can actually show that the conditional entropy is not only negative in this specific example, but for all pure entangled states:

Theorem 3.33 *Let $|\psi\rangle_{AB} \in \mathcal{H}_A \otimes \mathcal{H}_B$ be a pure bipartite state. $|\psi\rangle_{AB}$ is entangled if and only if $H(A|B)_\psi < 0$.*

Proof First, note that $H(AB)_\psi = 0$ because $|\psi\rangle_{AB}$ is a pure state. This reduces the problem to the following statement: Show that a pure state $|\psi\rangle_{AB}$ is entangled if and only if $H(B)_\psi = H(\text{Tr}_A(|\psi\rangle_{AB}\langle\psi|)) > 0$.

$\Rightarrow$ Suppose $|\psi\rangle_{AB}$ is entangled. This implies that its Schmidt rank d is strictly greater than one. Hence, according to (2.135), the marginal state of system B is

$$\rho_B = \text{Tr}_A(|\psi\rangle_{AB}\langle\psi|) = \sum_{i=1}^{d} \lambda_i^2 |i\rangle_B\langle i| \tag{3.84}$$

with $d > 1$ and eigenvalues λ_i^2. Since the sum consists of at least two terms, ρ_B cannot be a pure state and, therefore, $H(\rho_B) > 0$.

$\Leftarrow$ Suppose $H(B)_\psi > 0$. Consider the marginal state ρ_B as given in (3.84), but this time we do not know anything about the Schmidt rank d. By using the fact that the entropy is only a formula of the eigenvalues of ρ_B, i.e., the λ_i^2, we find that

$$H(\rho_B) = -\text{Tr}(\rho_B \log(\rho_B)) = -\sum_{i=1}^{d} \lambda_i^2 \log(\lambda_i^2). \tag{3.85}$$

If $d = 1$, there is only one eigenvalue $\lambda^2 = 1$ (because $\sum_i \lambda_i^2 = 1$), but this implies that $H(\rho_B) = 0$, which contradicts our assumption. Therefore, $d > 1$ and $|\psi\rangle_{AB}$ is entangled.

$\square$

Since the negativity of the conditional quantum entropy is such an important phenomenon in quantum information theory, there is a separate notion for the negativity of the conditional quantum entropy, which is called the *coherent information*:

---

[9]This is similar to the observation that Erwin Schrödinger made on entangled states in 1935 in his paper [11]: "The best possible knowledge of a *whole* does not necessarily include the best possible knowledge of all its *parts*, even though they may be entirely separated and therefore virtually capable of being 'best possibly known', i.e. of possessing, each of them, a representative of its own".

Definition 3.34 For a bipartite state $\rho_{AB} \in \mathcal{B}(\mathcal{H}_A \otimes \mathcal{H}_B)$ the coherent information $I(A\rangle B)_\rho$ is defined as

$$I(A\rangle B)_\rho = H(B)_\rho - H(AB)\rho. \tag{3.86}$$

The coherent information can be interpreted as a measure of quantum correlations that are present in a bipartite state ρ_{AB} shared by Alice and Bob. Although the coherent information is simply the negativity of the conditional entropy stated in Definition 3.31, it is useful in several areas of quantum information theory and therefore is treated as an information quantity in its own right. For example, it satisfies the following data processing inequality, which states that processing one system by some quantum channel $\mathcal{E}$ reduces quantum correlations:

Theorem 3.35 (Data Processing for Coherent Information) *Let $\rho_{AB} \in \mathcal{B}(\mathcal{H}_A \otimes \mathcal{H}_B)$ be a bipartite quantum state and $\mathcal{E} : \mathcal{B}(\mathcal{H}_B) \to \mathcal{B}(\mathcal{H}_{B'})$ be a quantum channel on Bob's system. Set $\sigma_{AB'} = \mathcal{E}_{B \to B'}(\rho_{AB})$. Then the following inequality holds*

$$I(A\rangle B)_\rho \geq I(A\rangle B')_\sigma. \tag{3.87}$$

The proof of this theorem can be found in [17, Thm. 11.9.3]. It directly implies the following relation for the conditional entropy:

$$H(A|B)_\rho \leq H(A|B')_\sigma. \tag{3.88}$$

In the case that $\mathcal{E}_{B \to B'}$ is a channel that represents a measurement on Bob's system and B' is the system after the measurement, this inequality states that measurements cannot decrease the conditional entropy.

Exercise 3.36 Calculate the coherent information of the maximally entangled state $|\Phi^+\rangle_{AB} = \frac{1}{\sqrt{2}}(|00\rangle + |11\rangle)$.

3.2.4 Quantum Mutual Information

The quantum generalization of the classical mutual information is straightforward:

Definition 3.37 For a bipartite state $\rho_{AB} \in \mathcal{B}(\mathcal{H}_A \otimes \mathcal{H}_B)$, the quantum mutual information $I(A:B)_\rho$ is defined as

$$I(A:B)_\rho = H(A)_\rho + H(B)_\rho - H(AB)_\rho \tag{3.89}$$

$$= H(A)_\rho - H(A|B)_\rho \tag{3.90}$$

$$= H(B)_\rho - H(B|A)_\rho. \tag{3.91}$$

Exercise 3.38 Show that the quantum mutual information is non-negative: $I(A : B) \geq 0$. *Hint: Use the fact that conditioning does not increase the entropy.*

Exercise 3.39 Calculate the mutual information of the maximally entangled state $|\Psi^+\rangle_{AB} = \frac{1}{\sqrt{2}}(|01\rangle + |10\rangle)$.

Exercise 3.40 Show the following upper bound on the mutual information of a classical–quantum system XB: $I(X : B) \leq \log(d_X)$, where d_X is the dimension of the classical system X.

3.2.5 Quantum Relative Entropy

Analogous to the classical case, we can formulate a quantum version of the relative entropy:

Definition 3.41 For two density operators $\rho, \sigma \in \mathcal{B}(\mathcal{H}_A \otimes \mathcal{H}_B)$, the quantum relative entropy $D(\rho\|\sigma)$ is defined as

$$D(\rho\|\sigma) = \mathrm{Tr}\left(\rho \log(\rho) - \rho \log(\sigma)\right). \tag{3.92}$$

As in the classical case, this quantity can be infinite. This is the case if the kernel of σ has non-trivial intersection with the support of ρ. The above definition can actually be generalized to the case where σ is not a density matrix but only a positive semi-definite operator, but this case is not of any interest to us and therefore we neglect it here.

Similar to the classical case (see Theorem 3.22), the quantum relative entropy is non-negative (which is also known as *Klein's inequality*):

Theorem 3.42 *For two density operators* $\rho, \sigma \in \mathcal{B}(\mathcal{H}_A \otimes \mathcal{H}_B)$, *the quantum relative entropy* $D(\rho\|\sigma)$ *is non-negative:*

$$D(\rho\|\sigma) \geq 0, \tag{3.93}$$

with equality if and only if $\rho = \sigma$.

The proof of the above theorem uses the respective spectral decompositions to reduce the quantum relative entropy to its classical counterpart and exploits the fact that the classical relative entropy is non-negative. The full proof can, for example, be found in [7, Thm. 11.7].

We can now derive a relation between the quantum mutual information and the quantum relative entropy:

Corollary 3.43 *For a bipartite state $\rho_{AB} \in \mathcal{B}(\mathcal{H}_A \otimes \mathcal{H}_B)$ with marginal states $\rho_A = \mathrm{Tr}_B(\rho_{AB})$ and $\rho_B = \mathrm{Tr}_A(\rho_{AB})$, the following relation holds*

$$I(A:B) = D(\rho_{AB} \| \rho_A \otimes \rho_B). \tag{3.94}$$

Proof First, note that $\log(\rho_A \otimes \rho_B) = \log(\rho_A) \otimes \mathbb{I}_B + \mathbb{I}_A \otimes \log(\rho_B)$. Using this relation, we can perform the following manipulations:

$$\mathrm{Tr}\big(\rho_{AB} \log(\rho_A \otimes \rho_B)\big) = \mathrm{Tr}\Big(\rho_{AB}\big(\log(\rho_A) \otimes \mathbb{I}_B + \mathbb{I}_A \otimes \log(\rho_B)\big)\Big) \tag{3.95}$$

$$= \mathrm{Tr}\big(\rho_{AB} \log(\rho_A) \otimes \mathbb{I}_B\big) + \mathrm{Tr}\big(\rho_{AB} \log(\rho_B) \otimes \mathbb{I}_A\big) \tag{3.96}$$

$$= \mathrm{Tr}_A\big(\mathrm{Tr}_B\big(\rho_{AB} \log(\rho_A) \otimes \mathbb{I}_B\big)\big) \tag{3.97}$$

$$+ \mathrm{Tr}_B\big(\mathrm{Tr}_A\big(\rho_{AB} \log(\rho_B) \otimes \mathbb{I}_A\big)\big) \tag{3.98}$$

$$= \mathrm{Tr}_A\big(\rho_A \log(\rho_A)\big) + \mathrm{Tr}_B\big(\rho_B \log(\rho_B)\big) \tag{3.99}$$

$$= -H(A)_\rho - H(B)_\rho. \tag{3.100}$$

This directly implies the following identities for the relative entropy:

$$D(\rho_{AB} \| \rho_A \otimes \rho_A) = \mathrm{Tr}(\rho_{AB} \log(\rho_{AB})) - \mathrm{Tr}(\rho_{AB} \log(\rho_A \otimes \rho_B)) \tag{3.101}$$

$$= -H(AB)_\rho + H(A)_\rho + H(B)_\rho \tag{3.102}$$

$$= I(A:B)_\rho, \tag{3.103}$$

which is the statement of the corollary.   □

Exercise 3.44 Show that $D(\rho_{AB} \| \mathbb{I}_A \otimes \rho_B) = -H(A|B)_\rho$.

Exercise 3.45 Use the non-negativity of the quantum relative entropy to show that for a quantum system A of dimension d_A, it holds that $H(A)_\rho \le \log d_A$.

A very important inequality in quantum information theory is the fact that the relative entropy decreases when applying a quantum channel to it, which was shown in [5]:

Theorem 3.46 *For two quantum states ρ, σ, the quantum relative entropy can only decrease under the action of a quantum channel, i.e., a completely positive, trace-preserving map $\mathcal{E}$:*

$$D(\mathcal{E}(\rho) \| \mathcal{E}(\sigma)) \le D(\rho \| \sigma). \tag{3.104}$$

By using the fact that the partial trace operation is a quantum channel (as discussed in Sect. 2.3), it directly follows that taking the partial trace of a bipartite system decreases the relative entropy, which is known as the *monotonicity of the quantum relative entropy*:

Theorem 3.47 *For two bipartite quantum states ρ_{AB} and σ_{AB} it holds that*

$$D(\rho_A \| \sigma_A) \leq D(\rho_{AB} \| \sigma_{AB}), \tag{3.105}$$

where $\rho_A = \mathrm{Tr}_B(\rho_{AB})$ and $\sigma_A = \mathrm{Tr}_B(\sigma_{AB})$.

3.2.6 One-Shot Entropies

The classical and quantum entropies we have discussed so far only apply in the so-called i.i.d. scenario, which stands for *independently* and *identically distributed*. This is because we need to repeat the corresponding experiment independently and infinitely many times in order to know the exact probabilities $p_X(x)$ of the outcomes. For the scenario where we only repeat the experiment a finite number of times (which is every practical implementation of a QKD experiment), we can use the so-called *one-shot entropies*, more precisely, the min- and max-entropies. These come from a family of entropies that is called *Rényi entropies* [10]. The classical counterparts to these entropies have not been discussed in this chapter since they are not of interest for these notes. An in-depth discussion of one-shot entropies can, for example, be found in [13].

We begin by introducing the quantum min-entropy as presented in [9]. We will encounter this entropy, for example, when we talk about *privacy amplification* in Sect. 4.2.3: Here, we need to transform Alice's classical bit string, which is correlated with Eve's quantum system, into a uniformly random string that is independent of Eve's knowledge.

As a motivation for the definition of the quantum min-entropy we consider an alternative derivation of the conditional quantum entropy $H(A|B)$, which was defined in Definition 3.31. First, we define

$$H(\rho_{AB}|\sigma_B) := -\mathrm{Tr}\left(\rho_{AB}\left(\log \rho_{AB} - \log \mathbb{I}_A \otimes \sigma_B\right)\right) \tag{3.106}$$

for some state $\sigma_B \in \mathcal{B}(\mathcal{H}_B)$. It can be rewritten as

$$H(\rho_{AB}|\sigma_B) = H(\rho_{AB}) - H(\rho_B) - D(\rho_B \| \sigma_B), \tag{3.107}$$

where $\rho_B = \mathrm{Tr}_A(\rho_{AB})$ and $D(\rho_B \| \sigma_B)$ as defined in Definition 3.41. Since the relative entropy cannot be negative (see Theorem 3.42), this expression takes its maximum for $\rho_B = \sigma_B$, for which $D(\rho_B \| \sigma_B)$ is zero. In this case, the above expression is equal to $H(A|B)$; hence, we can write

$$H(A|B) = \sup_{\sigma_B} H(\rho_{AB}|\sigma_B), \tag{3.108}$$

where the supremum ranges over all quantum states $\sigma_B \in \mathcal{B}(\mathcal{H}_B)$.

To define the quantum conditional min-entropy we use a similar approach: First, we define the min-entropy of $\rho_{AB} \in \mathcal{B}(\mathcal{H}_A \otimes \mathcal{H}_B)$ relative to a state $\sigma_B \in \mathcal{B}(\mathcal{H}_B)$ as

$$H_{\min}(\rho_{AB}|\sigma_B) = -\log \min\{\lambda : \rho_{AB} \leq \lambda \cdot \mathbb{I}_A \otimes \sigma_B\}. \tag{3.109}$$

Similar to the above construction, the quantum conditional min-entropy is then defined as the supremum over all states $\sigma_B \in \mathcal{B}(\mathcal{H}_B)$, which is formalized in the following definition:

Definition 3.48 The quantum conditional min-entropy of a state $\rho_{AB} \in \mathcal{B}(\mathcal{H}_A \otimes \mathcal{H}_B)$, conditioned on $\mathcal{H}_B$, is defined as

$$H_{\min}(A|B) = \sup_{\sigma_B} H_{\min}(\rho_{AB}|\sigma_B) \tag{3.110}$$

$$= -\log \min_{\sigma_B} \min\{\lambda : \rho_{AB} \leq \lambda \cdot \mathbb{I}_A \otimes \sigma_B\}, \tag{3.111}$$

where the minimum is over all states $\sigma_B \in \mathcal{B}(\mathcal{H}_B)$.

Although the definition is rather technical, the operational interpretation of the min-entropy exactly matches our needs for an application in privacy amplification: For classical–quantum states, the conditional min-entropy characterizes the amount of uniform randomness that we can extract from a classical random variable that is correlated with a quantum system such that the result is independent of the quantum system (see [9] and [4]).

The second one-shot entropy that we will need in our analysis of QKD protocols is the max-entropy:

Definition 3.49 The quantum conditional max-entropy of a state $\rho_{AB} \in \mathcal{B}(\mathcal{H}_A \otimes \mathcal{H}_B)$, conditioned on $\mathcal{H}_B$, is defined as

$$H_{\max}(A|B) = \max_{\sigma_B} \log \|\sqrt{\rho_{AB}}\sqrt{\mathbb{I}_A \otimes \sigma_B}\|_1^2. \tag{3.112}$$

A possible interpretation of the max-entropy is the following: Given a classical–quantum state ρ_{XB}, the max-entropy quantifies the size of the system that X can be compressed to, such that the original system can be recovered given access to the quantum system B (see [8]).

It is possible to modify the min- and max-entropies in order to account for errors and imperfections in the tasks that they characterize. The resulting entropies are called *smooth* min- and max-entropies. For these definitions, we first need to specify a region of states close to a fixed state, which is called an ϵ-ball.

Definition 3.50 For a subnormalized state $\rho \in S_\le(\mathcal{H})$ an ϵ-ball around the state ρ is defined as the set

$$\mathcal{B}^\epsilon(\rho) = \left\{ \tilde{\rho} : \tilde{\rho} \in S_\le(\mathcal{H}), P(\rho, \tilde{\rho}) \le \epsilon \right\}, \tag{3.113}$$

where $P(\rho, \tilde{\rho})$ is the purified distance that was defined in Definition 2.58.

Definition 3.51 For a quantum state $\rho_{AB} \in \mathcal{B}(\mathcal{H}_A \otimes \mathcal{H}_B)$ the smooth conditional min- and max-entropies are defined as

$$H_{\min}^\epsilon(A|B) = \max_{\rho' \in \mathcal{B}^\epsilon(\rho)} H_{\min}(A|B)_{\rho'} \tag{3.114}$$

$$H_{\max}^\epsilon(A|B) = \min_{\rho' \in \mathcal{B}^\epsilon(\rho)} H_{\max}(A|B)_{\rho'}. \tag{3.115}$$

The smooth min- and max-entropies fulfil a variety of interesting properties (see [13]). However, in these notes we only need two of them: The first one is a duality relation between min- and max-entropies, which was shown in [16]:

Theorem 3.52 *Given a pure state* $\rho_{ABC} \in \mathcal{B}(\mathcal{H})$ *and* $\epsilon \ge 0$, *then*

$$H_{\min}^\epsilon(A|B) = -H_{\max}^\epsilon(A|C). \tag{3.116}$$

The second important property of one-shot entropies is the *quantum asymptotic equipartition property*, which was shown in [15]:

Theorem 3.53 *Given a quantum state* $\rho_{AB} \in \mathcal{B}(\mathcal{H}_a \otimes \mathcal{H}_B)$, *then*

$$\lim_{\epsilon \to 0} \lim_{n \to \infty} \frac{1}{n} H_{\min}^\epsilon(A^n|B^n)_{\rho^{\otimes n}} = H(A|B) \tag{3.117}$$

$$\lim_{\epsilon \to 0} \lim_{n \to \infty} \frac{1}{n} H_{\max}^\epsilon(A^n|B^n)_{\rho^{\otimes n}} = H(A|B). \tag{3.118}$$

This theorem shows that in the limit of having an i.i.d. quantum state the min- and max-entropies both approach the conditional von Neumann entropy, which makes them generalizations of the von Neumann entropy to the one-shot scenario.

3.3   The Entropic Uncertainty Principle

Heisenberg's uncertainty relation for the momentum uncertainty and the position uncertainty of a particle is without doubt one of the most famous formulas in physics. However, from an information-theoretic view, it is not directly clear how to interpret the involved quantities: For instance, the uncertainty of position and momentum is formulated in terms of the standard deviation, which lacks an operational interpretation similar to the one we gave for the entropy. This makes it difficult to directly apply these kinds of uncertainty relations to information-theoretic tasks like the transmission of data over a noisy channel.

Additionally, and this is probably the most confusing point, the uncertainty principle seems to be defied if two parties share a maximally entangled state.[10] We can see this if we consider the maximally entangled bipartite state $|\Phi^+\rangle_{AB}$, which can be written in two different ways using either the computational or Hadamard basis:

$$|\Phi^+\rangle_{AB} = \frac{1}{\sqrt{2}} \left(|0\rangle_A |0\rangle_B + |1\rangle_A |1\rangle_B \right) \qquad (3.119)$$

$$= \frac{1}{\sqrt{2}} \left(|+\rangle_A |+\rangle_B + |-\rangle_A |-\rangle_B \right) . \qquad (3.120)$$

Suppose Alice measures the Z operator, that is, the operator $Z = |0\rangle_A \langle 0| - |1\rangle_A \langle 1|$ on her system. According to the formula in (3.119), Bob can then say with certainty which outcome Alice obtained by only accessing his part of the quantum state. Additionally, if Alice measures the operator $X = |+\rangle_A \langle +| - |-\rangle_A \langle -|$ on her system, according to (3.120) Bob can again guess her outcome with certainty. Bob has no uncertainty about the outcome of Alice's measurements, in spite of the fact that X and Z are incompatible observables.

The above phenomenon motivates a reformulation of the uncertainty principle that takes into account the possibility that Bob has access to a *quantum memory* correlated with Alice's system, where uncertainty is measured in terms of quantum entropy instead of the standard deviation. Suppose Alice and Bob's respective systems A and B are in the state $\rho_{AB} \in \mathcal{B}(\mathcal{H}_A \otimes \mathcal{H}_B)$. Alice then performs a measurement on her system that is modelled by the POVM $\{M_A^x\}$. After she has performed the measurement, the state of Alice and Bob's system is

$$\sigma_{XB} = \sum_x |x\rangle_A \langle x| \otimes \mathrm{Tr}_A \left(\left(M_A^x \otimes \mathbb{I}_B \right) \rho_{AB} \right), \qquad (3.121)$$

---

[10]This problem was already addressed by Einstein, Podolsky, and Rosen in their paper "Can Quantum-Mechanical Description of Physical Reality be Considered Complete?" [3] in 1935, shortly after quantum mechanics was established. From this observation, they concluded that quantum mechanics could not be complete.

where the measurement outcomes x are encoded into orthonormal states $\{|x\rangle_A\}$ of a classical register X. Bob's uncertainty about the outcome of the measurement can then be quantified by the conditional entropy $H(X|B)_\sigma$. In the same fashion, Alice can decide to perform a different measurement given by the POVM $\{N_A^y\}$. The state after the measurement is then given by

$$\tau_{YB} = \sum_y |y\rangle_A \langle y| \otimes \mathrm{Tr}_A \left(\left(N_A^y \otimes \mathbb{I}_B \right) \rho_{AB} \right). \tag{3.122}$$

Similar to the first case, Bob's uncertainty about the measurement outcome is given by $H(Y|B)_\tau$. We define Bob's total uncertainty about the measurement outcome to be the sum of the individual uncertainties: $H(X|B)_\sigma + H(Y|B)_\tau$. We also need a quantity that measures the incompatibility of the two POVMs $\{M_A^x\}$ and $\{N_A^y\}$. One way to quantify this is via the parameter

$$c = \max_{x,y} ||\sqrt{M_A^x}\sqrt{N_A^y}||_\infty^2, \tag{3.123}$$

where $|| \cdot ||_\infty$ is the infinity norm of an operator (which is simply the largest eigenvalue in the finite-dimensional case). This quantity is equal to one if the two measurements are maximally compatible. We can now state the uncertainty principle in the presence of quantum memory:

Theorem 3.54 *Suppose that Alice and Bob share a state $\rho_{AB} \in \mathcal{B}(\mathcal{H}_A \otimes \mathcal{H}_B)$ and Alice performs one of the POVMs $\{M_A^x\}$ and $\{N_A^y\}$. Then Bob's uncertainty of the measurement outcome can be lower bounded in the following way:*

$$H(X|B)_\sigma + H(Y|B)_\tau \geq \log \frac{1}{c} + H(A|B)_\rho, \tag{3.124}$$

with states σ_{XB} and τ_{YB} defined in (3.121) and (3.122), respectively, and the constant c as given in (3.123).

Proof The proof of this theorem can be found in [17, Thm. 11.9.5] or in [1], where it was originally introduced. ◻

The lower bound given above consists of two terms, one term that takes into account the incompatibility of the two measurements (which is independent of the state) and the other that relies only on the quantum state of the system. Since the conditional entropy $H(A|B)_\rho$ can become negative, as we have already seen, the bound can actually be lower than $\log(1/c)$. It might even be possible to reduce Bob's uncertainty about the measurement to zero by picking the right state!

Recall the scenario we have studied above: The system is in state $|\Phi^+\rangle_{AB}$ and Alice can perform either the Z or X measurement on her system. The POVM elements that correspond to the Z measurement are $\{|0\rangle\langle 0|, |1\rangle\langle 1|\}$, and the

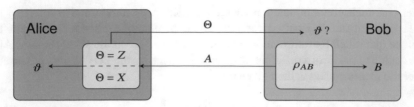

Fig. 3.6 Entropic uncertainty relation as a guessing game. Bob prepares the bipartite state ρ_{AB} and sends system A to Alice. Alice performs one of two measurements, X or Z, and sends her measurement choice Θ back to Bob. Bob then tries to guess Alice's measurement outcome ϑ using the side information he gets from his system B

POVM elements that correspond to the X measurement are $\{|+\rangle\langle+|, |-\rangle\langle-|\}$. The incompatibility of the two POVMs is then $c = \frac{1}{2}$, which implies $\log(1/c) = 1$. We have already seen in (3.83) that $H(A|B)_{\Phi^+} = -1$; hence,

$$H(X|B)_\sigma + H(Z|B)_\tau \geq 0, \tag{3.125}$$

which is consistent with the observation that Bob knows Alice's measurement outcome with certainty.

The above theorem can be reformulated as a guessing game (depicted in Fig. 3.6): Suppose Bob prepares a bipartite quantum state ρ_{AB}. He sends system A to Alice and keeps system B. Alice then performs either the X measurement or the Z measurement and sends the information about which measurement she performs back to Bob. Bob's task is then, by using his quantum system B, to guess Alice's measurement outcome. If Bob has prepared the maximally entangled state $|\Phi^+\rangle_{AB}$, he is able to predict Alice's measurement outcome with certainty. However, in general, he has some uncertainty about the outcome, which is lower bounded by the quantity in Theorem 3.54.

It is furthermore possible to formulate an entropic uncertainty relation in the one-shot setting using the min- and max-entropies introduced in the previous section. In [14], the authors show the following:

Theorem 3.55 *Let $\rho_{ABE} \in \mathcal{B}(\mathcal{H}_A \otimes \mathcal{H}_B \otimes \mathcal{H}_E)$ and $\epsilon \geq 0$. Define two POVMs with elements $\{M_A^x\}$ and $\{N_A^y\}$ that are acting on system A, resulting in outcomes X and Z. Then*

$$H_{\min}^\epsilon(X|E) + H_{\max}^\epsilon(Z|B) \geq \log\frac{1}{c}, \tag{3.126}$$

with $c = \max_{x,y} \|\sqrt{M_A^x}\sqrt{N_A^y}\|_\infty^2$.

This uncertainty relation is useful for quantum cryptography, since it lower bounds the amount of information an adversary E has on the outcome of Alice's measurement X. We go into more detail about the use of entropic uncertainty relations in quantum key distribution in Sect. 5.5.1.

References

1. Berta, M., Christandl, M., Colbeck, R., Renes, J.M., Renner, R.: The uncertainty principle in the presence of quantum memory. Nat. Phys. **6**(9), 659–662 (2010). https://doi.org/10.1038/nphys1734
2. Cover, T.M., Thomas, J.A.: Elements of Information Theory. Wiley, New York (2006)
3. Einstein, A., Podolsky, B., Rosen, N.: Can quantum-mechanical description of physical reality be considered complete? Phys. Rev. **47**(10), 777–780 (1935). https://doi.org/10.1103/physrev.47.777
4. Konig, R., Renner, R., Schaffner, C.: The operational meaning of min- and max-entropy. IEEE Trans. Inform. Theory **55**(9), 4337–4347 (2009). https://doi.org/10.1109/tit.2009.2025545
5. Lindblad, G.: Completely positive maps and entropy inequalities. Commun. Math. Phys. **40**(2), 147–151 (1975). https://doi.org/10.1007/bf01609396
6. MacKay, D.J.C.: Information Theory, Inference and Learning Algorithms. Cambridge University Press, Cambridge (2003)
7. Nielsen, M.A., Chuang, I.L.: Quantum Computation and Quantum Information. Cambridge University Press, Cambridge (2000)
8. Renes, J.M., Renner, R.: One-shot classical data compression with quantum side information and the distillation of common randomness or secret keys. IEEE Trans. Inform. Theory **58**(3), 1985–1991 (2012). https://doi.org/10.1109/tit.2011.2177589
9. Renner, R.: Security of quantum key distribution. Ph.D. thesis, ETH Zurich (2005). https://doi.org/10.3929/ethz-a-005115027
10. Rényi, A.: On measures of information and entropy. In: Proceedings of the fourth Berkeley Symposium on Mathematics, Statistics and Probability, vol. 1, pp. 547–561 (1961)
11. Schrödinger, E.: Discussion of probability relations between separated systems. Math. Proc. Cambridge Philos. Soc. **31**(4), 555–563 (1935). https://doi.org/10.1017/s0305004100013554
12. Shannon, C.E.: A mathematical theory of communication. Bell Syst. Tech. J. **27**(3), 379–423 (1948). https://doi.org/10.1002/j.1538-7305.1948.tb01338.x
13. Tomamichel, M.: A framework for non-asymptotic quantum information theory. Ph.D. thesis, ETH Zurich (2012). https://doi.org/10.3929/ETHZ-A-7356080
14. Tomamichel, M., Renner, R.: Uncertainty relation for smooth entropies. Phys. Rev. Lett. **106**(11), 110506 (2011). https://doi.org/10.1103/physrevlett.106.110506
15. Tomamichel, M., Colbeck, R., Renner, R.: A fully quantum asymptotic equipartition property. IEEE Trans. Inform. Theory **55**(12), 5840–5847 (2009). https://doi.org/10.1109/tit.2009.2032797
16. Tomamichel, M., Colbeck, R., Renner, R.: Duality between smooth min- and max-entropies. IEEE Trans. Inform. Theory **56**(9), 4674–4681 (2010). https://doi.org/10.1109/tit.2010.2054130
17. Wilde, M.M.: Quantum Information Theory, 2nd edn. Cambridge University Press, Cambridge (2017). https://doi.org/10.1017/9781316809976

Quantum Key Distribution Protocols

4

Abstract

The goal of any quantum key distribution (QKD) protocol is to generate a shared secret key between two distant parties over a public communication channel. The crucial point here is that the key generating protocol is *provably* secure against any possible attack that an eavesdropper can perform. It is the law of physics (or, in fact, quantum mechanics) that guarantees the security of the protocol, not only the technical limitations that exist in practical implementations. Therefore, one can be sure that the protocol will be secure until eternity, and not only until someone invents a crazily powerful decryption machine (or, to be precise, the protocol will be secure as long as quantum mechanics is not disproved). In general, a quantum key distribution protocol can be divided into two parts: The first part is the quantum transmission phase, in which Alice and Bob send and/or measure quantum states. The second part is the classical post-processing phase, where they turn the bit strings generated in the quantum phase into a pair of secure keys.

Accessible introductions to the area of quantum cryptography can be found in [5] and [8]. Both of these books also include some chapters about implementations of quantum communication protocols, which is a topic we do not cover here. Furthermore, [24] contain a section about quantum cryptography that discusses the BB84 protocol and its security in detail. An overview of the concepts of quantum cryptography can be found in [17] and a review of the more recent developments in this area is given in [25].

4.1   Quantum Transmission

We begin by discussing the quantum phase of a QKD protocol. So far, we have only seen a single protocol, the BB84 protocol in Sect. 1.3. This is an example of a *prepare-and-measure protocol*, where quantum states are prepared, sent via a quantum channel, and measured afterwards. There is a second class of protocols, the so-called *entanglement-based protocols*, where Alice and Bob hold pairs of entangled states and perform measurements on their respective subsystems to generate the raw key. In contrast to the first type of protocol, these kinds of protocols do not require the sending of quantum states, i.e., no quantum channels are involved. They do, however, require a source that provides entangled states for Alice and Bob.

We will first explain the individual steps in each of these protocol types (mostly by looking at examples) and then show how every prepare-and-measure protocol is equivalent to an entanglement-based protocol. This equivalence is extremely helpful when it comes to security proofs, since entanglement-based protocols are in general easier to analyse because they do not involve quantum channels.

4.1.1   Prepare-and-Measure Protocols

We again return to the BB84 protocol to study how prepare-and-measure protocols work. When we met the protocol the first time in Sect. 1.3 we simply considered the polarization of photons as a way to implement qubits. After we have introduced the necessary tools in the previous sections we can now actually analyse it mathematically.

In Fig. 4.1, the general schematic of a prepare-and-measure protocol is depicted. Similar to the depiction of the BB84 setting in Fig. 1.3, the two parties Alice and Bob have access to two channels: First, they have access to a quantum channel that Alice can use to send the quantum states she has prepared over to Bob. We assume that there are no restrictions (other than the laws of physics) on how Eve can interact with the messages sent over the quantum channel. The second channel is an authenticated classical channel where Alice and Bob can send classical messages

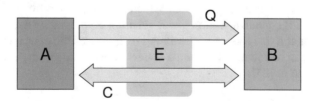

Fig. 4.1 The setting of a prepare-and-measure protocol. Alice (A) and Bob (B) have access to a quantum channel (Q), where Alice can send quantum states to Bob. Eve (E) can tap into this channel and interact with the quantum states without restrictions. They also have access to an authenticated classical channel (C), where they can send classical messages back and forth. Here, Eve can listen to the classical messages, but she cannot change them

back and forth. The fact that the channel is authenticated means that Alice and Bob can be sure they are communicating with each other. This excludes scenarios where Eve, for example, has broken into Bob's lab and simply replaced him. In this case, it would be pointless to try to produce a secret key. Furthermore, while Eve can listen to all the communication that is sent over the classical channel, she cannot change the classical messages.

In the following, we discuss the steps of a prepare-and-measure protocol by studying some examples. We focus here on the quantum transmission phase and do not go into any detail about the classical post-processing steps. These will be discussed later in Sect. 4.2.

4.1.1.1 The BB84 Protocol

In the BB84 protocol, Alice starts by choosing two random classical bit strings $a = (a_1, a_2, \ldots, a_{4n})$ and $b = (b_1, b_2, \ldots, b_{4n})$, both of length $4n$. The first string determines the bit value she wants to send (0 or 1), and the second string determines the basis she uses to encode the bit: 0 represents computational basis (also called the Z-basis) and 1 represents the Hadamard basis (or X-basis). The measurement bases are also depicted on the left side of Fig. 4.2. According to the strings a and b, Alice prepares a block of $4n$ qubits

$$|\psi\rangle_A = \bigotimes_{i=1}^{4n} |\psi_{a_i b_i}\rangle_A, \tag{4.1}$$

where a_i is the i-th bit of string a and b_i is the i-th bit of string b: Hence, each of the individual qubits is in one of the four states

$$|\psi_{00}\rangle_A = |0\rangle_A \tag{4.2}$$

$$|\psi_{10}\rangle_A = |1\rangle_A \tag{4.3}$$

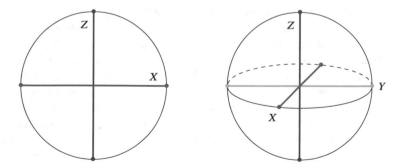

Fig. 4.2 Measurement directions for the BB84 and six-state protocols. On the left side, the two measurement bases (Z and X) for the BB84 protocol are depicted. It shows that they span the x-z-plane of the Bloch sphere. On the right side, the measurement directions for the six-state protocol (X, Y, and Z) are depicted. They span the whole Bloch sphere

$$|\psi_{01}\rangle_A = \frac{1}{\sqrt{2}} (|0\rangle_A + |1\rangle_A) \tag{4.4}$$

$$|\psi_{11}\rangle_A = \frac{1}{\sqrt{2}} (|0\rangle_A - |1\rangle_A). \tag{4.5}$$

Note that the four states are not all mutually orthogonal: For instance, $_A\langle\psi_{00}|\psi_{01}\rangle_A = \frac{1}{\sqrt{2}}$. This property ensures that there is no measurement that can perfectly distinguish between all of the states with certainty (which would make Eve's job pretty easy). Alice then sends the state $|\psi\rangle_A$ to Bob over the quantum channel, which we denote $\mathcal{E}$.

Bob receives a state $\mathcal{E}(|\psi\rangle_A\langle\psi|)$ and publicly announces this fact. $\mathcal{E}$ describes both the effect of applying the channel and Eve's interaction with the state. Since Eve has no knowledge about string b, which determines the basis that was used for encryption, she can only guess. If her guess was wrong, she would not gain any information about the bit Alice encoded (i.e., about the corresponding bit from string a). Moreover, she would disturb the state Bob receives.

Bob also does not have any knowledge about b, so he, as well, has to guess in which basis he should measure the state he has received. Hence, for each qubit he decides randomly if he measures it in the computational basis or in the Hadamard basis. As a result, he then holds a string a' that represents the decoded bit values and a string b' where his choice of basis is stored, both of length $4n$.

After this step, the *sifting* step follows. Alice publicly announces her choice of basis, i.e., the string b, via the classical communication channel. Bob compares his string b' to Alice's string b and tells Alice at which positions i their strings differ. They then discard all the pairs $\{a_i, a'_i\}$ for which $b_i \neq b'_i$. The remaining bits in the strings a and a' satisfy $a_i = a'_i$ (in the ideal case), since for these bits Bob has measured in the same basis that Alice has prepared the qubits in. Note that the announcement of string b does not reveal any information about a or a'. However, it is important that Alice does not publish b before Bob has announced the reception of the state $\mathcal{E}(|\psi\rangle_A\langle\psi|)$. Since Eve also gets the information about b, she otherwise would have known exactly in which basis to measure the qubits in order to not change the states Bob receives. After this step, Alice and Bob hold strings a and a' each of length about $2n$, since the probability that Bob chooses the wrong basis is $\frac{1}{2}$.

The quantum transmission phase is now over. The remaining steps are part of the classical post-processing: First, Alice and Bob first estimate how much noise or eavesdropping has happened during their communication, which is called *parameter estimation*. Suppose they use half of their bits, i.e., after the parameter estimation step, n key bits remain. They then perform *error correction* and *privacy amplification* to turn their partially secret, correlated strings a and a', each of length n, into a pair of identical, secure keys of length $m < n$. The details of these steps are studied in Sect. 4.2.

The steps of the BB84 protocol are summarized in the following:

BB84 Protocol

1. Alice chooses two random classical bit strings $a = (a_1, \ldots a_{4n})$ and $b = (b_1, \ldots b_{4n})$.
2. She encodes the bits in a into the computational basis if the corresponding bit in b is 0, and into the Hadamard basis if the bit in b is 1, which results in a $4n$ block of qubits as given in (4.1).
3. Bob receives the $4n$ qubits and publicly announces this fact.
4. Bob measures each qubit in the computational or Hadamard basis at random that yields strings a' and b'.
5. Alice announces b.
6. Alice and Bob discard any bits a_i, a_i' for which $b_i \neq b_i'$. After this step, there are about $2n$ bits left.
7. Alice and Bob use n bits to estimate the amount of errors (and therefore the information that Eve has on the bit strings).
8. They perform error correction and privacy amplification on the remaining strings of n bits to obtain a shared secret key of m bits.

Exercise 4.1 Consider a setting where Alice and Bob use a noiseless quantum channel to perform the BB84 protocol and no eavesdropping is happening. Suppose a_i' is Bob's measurement result of measuring the qubit $|\psi_{a_i b_i}\rangle$. Show that if $b_i' \neq b_i$, i.e., in the case that Bob has chosen a different measurement basis than Alice, his result a_i' is random and completely unrelated with Alice's value a_i. On the other hand, show that if $b_i' = b_i$, then $a_i' = a_i$.

A Simple Eavesdropping Strategy for the BB84 Protocol

Let us have a closer look at how eavesdropping interferes with the states and how Alice and Bob can detect it. We will sketch a very simple strategy that Eve can pursue, which is the *intercept-and-resend* strategy. Here, Eve intercepts all $4n$ qubits that Alice sends to Bob. Since cloning the states is forbidden by the no-cloning theorem (see Sect. 2.4), the simplest way to get information about the states is to measure them. At this point of the protocol, Alice has not yet announced her choice of basis, so Eve has to guess in which basis she has to measure the states, thereby randomly choosing the basis she measures in (either the computational (C) or Hadamard (H) basis). In about half of the cases, i.e., for $2n$ of the qubits, her basis will be the same as the one that Alice chose for encoding the bit and Eve gets completely correlated bit values. In the other $2n$ cases, however, her guess will be

Table 4.1 Simple eavesdropping strategy for the BB84 protocol. Eve intercepts, measures, and resends every qubit that Alice sends. This introduces errors (highlighted in grey) to Alice and Bob's key bits that they can exploit to detect the eavesdropping during the parameter estimation (P.E.) step

Key bit	0	1	1	1	0	0	1	1	1	0
Alice's basis	C	C	H	C	C	H	H	C	H	H
Alice's state	$\|\psi_{00}\rangle$	$\|\psi_{10}\rangle$	$\|\psi_{11}\rangle$	$\|\psi_{10}\rangle$	$\|\psi_{00}\rangle$	$\|\psi_{01}\rangle$	$\|\psi_{11}\rangle$	$\|\psi_{10}\rangle$	$\|\psi_{11}\rangle$	$\|\psi_{01}\rangle$
Eve's basis	H	C	C	H	C	H	C	H	C	H
Eve's state	$\|\psi_{01}\rangle$	$\|\psi_{10}\rangle$	$\|\psi_{10}\rangle$	$\|\psi_{01}\rangle$	$\|\psi_{00}\rangle$	$\|\psi_{11}\rangle$	$\|\psi_{10}\rangle$	$\|\psi_{11}\rangle$	$\|\psi_{10}\rangle$	$\|\psi_{01}\rangle$
Bob's basis	H	C	H	H	C	C	H	C	H	C
Bob's result	0	1	0	0	0	0	1	1	0	0
Sifting	✗	✓	✓	✗	✓	✗	✓	✓	✓	✗
Key bit		1	0		0		1	1	0	
P.E.			✗				✓	✓		

wrong and she gets a random result.[1] Of course, as long as Alice has not revealed her choice of basis, Eve does not know which states she has measured in the correct basis.

Since Bob is expecting to receive the qubits from Alice, Eve has to send states to Bob. Because she does not know which are the correct bases, she simply prepares each qubit in the same basis that she has used for the measurement. Hence, $2n$ of the qubits will be prepared in the wrong basis. Bob then receives the qubits and measures them, again by randomly choosing the measurement basis. In n cases Bob and Alice have chosen the same basis, but Eve's basis is different. Since in these cases Bob gets a random result, there will be $\frac{n}{2}$ errors in the sifted key. Since the length of the sifted key is $2n$, this corresponds to an error rate of 25%. Hence, if Alice and Bob observe such a high error rate during parameter estimation they abort the protocol.

We have depicted an example of this eavesdropping attack using ten qubits in Table 4.1. After the sifting step, Alice and Bob are left with six of the initially ten qubits. Next, Alice and Bob use half of the remaining bits to estimate the knowledge that Eve has. They find that one of those three bits is wrong; hence, they have an estimated error of $\frac{1}{3}$ in their key bit string. Since this error is above 25%, they know that Eve has intercepted the communication and abort the protocol.

What about Eve's knowledge? How much information on the raw key (i.e., the key string after the parameter estimation step) did she get? Since she guesses the correct basis in $\frac{1}{2}$ of the cases, she knows 50% of the key bits. Let us check how

---

[1]For instance, suppose Alice has prepared the state $\|\psi_{00}\rangle$, which is in the computational basis. If Eve measures this state in the Hadamard basis, she gets the result 0 50% of the time and the result 1 also 50% of the time.

much knowledge of the key bits she has in the example discussed in Table 4.1. After she has measured the qubits, her bit string is

$$0 \quad 1 \quad 1 \quad 0 \quad 0 \quad 1 \quad 1 \quad 1 \quad 1 \quad 0$$

After Alice announces her choice of basis, Eve discards those bits where she has measured in a different basis than Alice, so she is left with

$$1 \qquad \qquad 0 \quad 1 \qquad \qquad 0$$

She also has to discard those bits that Alice and Bob discard during their sifting procedure:

$$1 \qquad \qquad 0$$

At last, she has to discard the bits that were used for parameter estimation, which leaves her with

$$0$$

Therefore, in the end she knows one of the three bits of the raw key. Of course, since we have only used 10 qubits in this example, the numbers we get for the error fraction and the amount of Eve's knowledge are not very meaningful, but still you get the idea.

Exercise 4.2 A QKD protocol that is even simpler than the BB84 protocol is the *B92 protocol*, which was presented by Charles Bennett in 1992 [4]. Here, Alice prepares her qubits in one of the two different states (instead of four). She first generates a bit string a and for each bit a_i she then prepares the i-th qubit in the state

$$|\psi\rangle = \begin{cases} |0\rangle & \text{if } a_i = 0 \\ |+\rangle = \frac{|0\rangle + |1\rangle}{\sqrt{2}} & \text{if } a_i = 1. \end{cases} \tag{4.6}$$

Alice sends her qubits to Bob, who generates a random bit string a', which determines the basis in which he measures the qubits: If $a'_i = 0$, he measures them in the Z-basis, and if $a'_i = 1$, he chooses the X-basis. The measurement results form a second bit string b: if Bob obtains the result -1 then i-th bit of b is $b_i = 0$, and if he obtains $+1$, then $b_i = 1$. Bob then announces the bit string b (but keeps a' secret), and they keep only those pairs $\{a_i, a'_i\}$ for which $b_i = 1$.

1. Show that $b_i = 0$ when $a_i = a'_i$. Hence, $b_i = 1$ only if $a_i = a'_i \oplus 1$ (where $\oplus$ denotes addition modulo 2).

2. For the i-th qubit, what is the probability that Bob will obtain $b_i = 1$? *Hint: Note that Alice and Bob choose the strings a and a' at random, hence here each bit has probability $1/2$ to be 0 and probability $1/2$ to be 1.*
3. After discarding those bits for which $b_i = 0$, how can Alice and Bob form their private keys?

The B92 protocol shows that the impossibility of perfectly distinguishing two non-orthogonal quantum states lies at the heart of quantum key distribution. Similar to the BB84 protocol, it enables Alice and Bob to establish a secret key because it is impossible for an eavesdropper to distinguish between states without disturbing the correlation between Alice's and Bob's bits.

4.1.1.2 The Six-State Protocol

The six-state protocol is a variation of the BB84 protocol that was proposed in [7] and [2]. It uses three pairs of orthogonal states (i.e., six states in total) instead of two. Additionally to the states that are used in the BB84 protocol, here we also include the eigenstates of the Y measurement operator, which are

$$|\psi_{y+}\rangle = \frac{1}{\sqrt{2}} \begin{pmatrix} 1 \\ i \end{pmatrix}, \qquad |\psi_{y-}\rangle = \frac{1}{\sqrt{2}} \begin{pmatrix} 1 \\ -i \end{pmatrix}. \tag{4.7}$$

Therefore, we now have three mutually unbiased bases. The protocol follows the same steps as the BB84 protocol with only minor adjustments: Alice now chooses randomly between three different encoding bases. This has the effect that in the sifting phase, where Alice and Bob compare their respective bases and discard the bits when encoding and measurement basis do not match, now approximately $2/3$ of the bits are discarded instead of $1/2$.

Although the fraction of discarded bits is higher in this protocol, it has the advantage that the eavesdropper has less knowledge, since he now also has to guess between three different bases. Intuitively speaking, this is due to the fact that the six states now span the whole Bloch sphere instead of only the circle that is spanned by the four states used in the BB84 protocol. The two settings are compared in Fig. 4.2. For a given disturbance of the eavesdropper, one can show that the six-state protocol achieves a higher secret key rate than the BB84 protocol.

4.1.1.3 The SARG04 Protocol

The SARG04 protocol is a variation of the BB84 protocol introduced in [28][2] that is tailored to be robust against the so-called photon-number-splitting (PNS) attacks. This is an attack that exploits weaknesses in a common experimental implementation of QKD protocols. It is convenient to implement qubits using the polarization degree of freedom of photons, as we have seen in Sect. 1.3. In the ideal case each qubit is represented by one photon. However, in practice, ideal single-

---

[2]There is also an entanglement-based version of this protocol [6].

photon sources do not exist. Therefore, one often uses weak laser pulses to encode the bits. In this implementation, it is possible that photons are produced in multi-photon bunches, so Eve can perform an attack where she keeps one of the photons and lets the rest pass to Bob. She then just needs to wait until Alice announces her choice of basis, so she can measure the stored photons in the correct basis, hence receiving perfect knowledge about the key. We discuss this attack in more detail in Sect. 5.3.

The SARG04 protocol addresses this problem by using a different sifting procedure. The sifting step is the only step that differs in the SARG04 protocol compared to the BB84 protocol. Instead of publicly announcing the bases Alice uses to encode the bits, Alice and Bob pursue the following strategy: For every qubit she has sent, Alice chooses one state from the computational basis and one from the Hadamard basis in a way that the actual state of the qubit is one of these states. She then publicly announces the two states and notes (privately) which of the two states is the right one. For example, she notes 0 if the actual qubit state was the computational basis state and 1 if it was the Hadamard basis state. This piece of information is the secret key bit she wishes to communicate to Bob.

In order to obtain the secret bit, Bob must be able to distinguish between the two candidate states. He knows that the qubit state he received was one of the two candidate states that Alice has announced. On the basis of his measurement, he checks which test state his result is consistent with. If it is consistent with both of the states, he announces the bit to be invalid, since he is unable to determine which of the states was the one that was transmitted from his measurement outcome. If, on the other hand, one of the test states is inconsistent with his observed measurement outcome, Bob can retrieve the secret key bit and announces the bit to be valid.

Let us illustrate this scheme with an example. Recall that the state Alice sends is one of the four Bell states $|\psi_{00}\rangle$, $|\psi_{10}\rangle$, $|\psi_{01}\rangle$, $|\psi_{11}\rangle$ defined in (4.2)–(4.5). Suppose the state she has sent is $|\psi_{00}\rangle$. In the sifting step, she chooses to announce the states $|\psi_{00}\rangle$ and $|\psi_{01}\rangle$ and notes 0 as the secret bit. Depending on the basis Bob chose for his measurement, this can either be a valid or an invalid bit:

1. Suppose Bob chose to measure the state in the computational basis. The only possible outcome for this measurement is $|\psi_{00}\rangle$. This outcome is clearly consistent with the candidate state $|\psi_{00}\rangle$. However, this outcome is also possible if the transmitted state had been $|\psi_{01}\rangle$, since measuring this state in the computational basis yields $|\psi_{00}\rangle$ and $|\psi_{10}\rangle$ each with probability $\frac{1}{2}$. Therefore, Bob announces the bit to be invalid.

2. Suppose Bob has measured the state in the Hadamard basis. In this case, he obtains either $|\psi_{01}\rangle$ or $|\psi_{11}\rangle$, each with a probability of $\frac{1}{2}$. If his outcome is $|\psi_{01}\rangle$, this is again consistent with both of the candidate states. On the other hand, if his outcome is $|\psi_{11}\rangle$, then he can be certain that the state Alice has sent is $|\psi_{00}\rangle$, since this outcome can never be obtained from the state $|\psi_{01}\rangle$. Thus, in this case Bob is able to retrieve the secret bit, namely 0, and he announces that the bit is valid.

After the sifting step, Alice and Bob are left with roughly $\frac{1}{4}$ of the initial bits instead of the $\frac{1}{2}$ in the BB84 protocol. However, it is shown that the SARG04 protocol is provably better against PNS attacks than the BB84 protocol. The advantage of this protocol is that Alice never reveals her encoding bases. As a result, Eve has to store more photons to get reliable information about the secret bits, which raises the chances that her attack is detected.

4.1.2 Entanglement-Based Protocols

In order to establish a secret key between Alice and Bob, we can also make use of entanglement. For instance, with regard to the PNS attack it might be beneficial to use entangled photons since the likelihood of simultaneously producing two entangled photon pairs is very low; hence, the PNS attack is much less effective. In general, the setting of an entanglement-based protocol, which is depicted in Fig. 4.3, includes a source that distributes entangled states between Alice and Bob. There are no conditions on where this source is or who controls it. It can be in Alice's lab and she distributes the entangled qubit pairs before performing the steps of the protocol, or some third party (which is usually called Charlie) distributes the pairs between Alice and Bob. It is even possible that Eve is in control of the source. Therefore, we regard the source as a completely untrusted device and, to account for the worst case, we usually assume that it is Eve who has perfect control of the source. As in the previous scenario, Alice and Bob have access to an authenticated classical channel where Eve can listen to the communication but cannot change the messages. Entanglement-based protocols can be easier to analyse in terms of security, since there is no quantum channel between Alice and Bob that has to be taken into account.

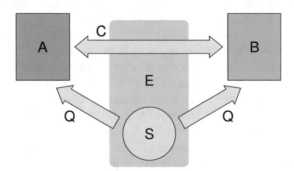

Fig. 4.3 The setting of an entanglement-based protocol. Alice (A) and Bob (B) have access to an authenticated classical channel (C) that they can use to send classical messages back and forth. Eve (E) can listen to (but not change) all communication over the classical channel. The scheme furthermore includes a source (S) that provides entangled states for Alice and Bob via quantum channels (Q). We assume that Eve has total control over this source to account for the worst case scenario

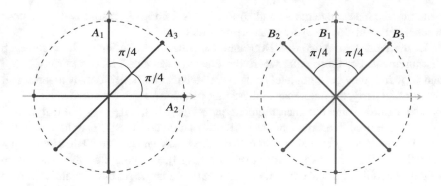

Fig. 4.4 Measurement directions for the Ekert protocol. The measurements are depicted in the x-z-plane of the Bloch sphere. On the left side are the three different measurements that Alice can choose between, and on the right side Bob's possible measurement directions are shown

4.1.2.1 The Ekert Protocol

In 1991, Arthur Ekert developed a scheme that exploits entanglement to generate a secret key [16]. Here, we follow the description of the protocol given in [9]. The protocol works as follows: Alice and Bob have access to a source that distributes maximally entangled pairs of qubits among them, for instance, states of the form

$$|\Psi^-\rangle_{AB} = \frac{1}{\sqrt{2}} \left(|01\rangle_{AB} - |10\rangle_{AB}\right). \tag{4.8}$$

For each of these bipartite states $|\Psi^-\rangle_{AB}$ Alice and Bob measure an observable that is randomly chosen from the sets $\{A_i\}$ and $\{B_i\}$, respectively. These observables are spin components lying in the x-z-plane of the Bloch sphere and are depicted in Fig. 4.4. In general, these operators are defined as

$$A_i = \cos \varphi_i^A + \sin \varphi_i^A,$$

$$B_i = \cos \varphi_i^B + \sin \varphi_i^B$$

with $\varphi_1^A = 0$, $\varphi_2^A = \frac{\pi}{2}$, and $\varphi_3^A = \frac{\pi}{4}$ for Alice and $\varphi_1^B = 0$, $\varphi_2^B = -\frac{\pi}{4}$, and $\varphi_3^B = \frac{\pi}{4}$ for Bob. In terms of the measurement operators $Z = |0\rangle\langle 0| - |1\rangle\langle 1|$ and $X = |+\rangle\langle +| - |-\rangle\langle -|$, the measurements can also be written as

$$A_1 = Z, \qquad\qquad B_1 = Z,$$

$$A_2 = X, \qquad\qquad B_2 = \frac{1}{\sqrt{2}} (Z - X),$$

$$A_3 = \frac{1}{\sqrt{2}} (Z + X), \qquad\qquad B_3 = \frac{1}{\sqrt{2}} (Z + X).$$

Note that the measurements A_1 and B_1 as well as A_3 and B_3, respectively, are those where Alice and Bob measure in the same direction.

In the next step, Alice and Bob announce the directions they chose for each measurement. For those pairs where the directions match, i.e., the pairs (A_1, B_1) and (A_3, B_3), they get completely anti-correlated results. Therefore, by inverting all bits for one party, the outcomes of these measurements form the sifted key.

The results from the measurement pairs (A_1, B_3), (A_1, B_2), (A_2, B_3), and (A_2, B_2) are used to estimate how much information an eavesdropper has about the key. This is done by checking a so-called *CHSH inequality*. The CHSH inequality, named after the initials of its four discoverers [13], is a bound on the expectation values of certain classical correlations. It is part of a larger set of inequalities known as Bell inequalities (because the first one was found by John Bell [3]). Suppose you have four classical random variables, A_1, A_2, B_2, B_3. Suppose each of them can take one of two values, $+1$ or -1. One can easily verify that $A_1(B_3 + B_2) + A_2(B_3 - B_2) = \pm 2$, simply by checking all possibilities. By taking the expectation value of these quantities over N assignments of the random variables, we get

$$|\langle A_1(B_3 + B_2) + A_2(B_3 - B_2)\rangle| \leq 2.$$

Since taking the expectation of a random variable is a linear operation, we can rewrite this and obtain the CHSH inequality:

$$S := |\langle A_1 B_3\rangle + \langle A_1 B_2\rangle + \langle A_2 B_3\rangle - \langle A_2 B_2\rangle| \leq 2,$$

where $\langle A_i B_j\rangle = \frac{1}{N} \sum A_i^\nu B_j^\nu$, and A_i^ν and B_j^ν represent the assigned values ν to the random variables A_i and B_i.

We can now consider A_1, A_2, B_2, B_3 to be quantum observables as described in the Ekert protocol. The expectation value for their products is then given by

$$\langle A_i B_j\rangle = \mathrm{Tr}\left(A_i \otimes B_j \rho\right).$$

Using the measurement directions defined in the Ekert protocol, we can evaluate their expectation values with respect to the state $\rho = |\Psi^-\rangle\langle\Psi^-|$. For instance, the expectation value of A_1 and B_3 is

$$\langle A_1 B_3\rangle = \langle\Psi^-|\left(Z \otimes \frac{1}{\sqrt{2}}(Z + X)\right)|\Psi^-\rangle = -\frac{1}{\sqrt{2}}.$$

In this way we can evaluate all terms in the sum of expectation values S and find that

$$S = 2\sqrt{2}.$$

This is a violation of the CHSH inequality that we have derived above and tells Alice and Bob that they share a maximally entangled state. In this case, Eve has no information about the key, since a maximally entangled bipartite state cannot be entangled with a third party.

This is actually the highest value for S that can be achieved. In general, it is possible to obtain lower values for S that still violate the CHSH inequality. In this case, Eve can have some knowledge about the key. However, it is still possible to extract a secret key from this data as long as there is some violation of the CHSH inequality. If $S \leq 2$, this indicates that Alice and Bob share a pair of separable states, i.e., it is impossible to generate a secret key (which was shown in [14]).

If their measurement results pass the test, Alice and Bob can proceed to the next step of the protocol and obtain the final secret key by doing error correction and privacy amplification. The steps of the protocol are summarized below:

Ekert91

1. Alice and Bob distribute a number of $|\Psi^-\rangle_{AB}$ states (4.8) between them, where the first subsystem belongs to Alice and the second one to Bob.
2. For each state, Alice and Bob randomly choose a measurement from the sets $\{A_i\}$ and $\{B_i\}$, respectively.
3. Alice and Bob announce the bases they chose for each measurement. In the cases where the directions match, (A_1, B_1) and (A_3, B_3), the results form the sifted key.
4. The results where Alice and Bob chose the directions (A_1, B_3), (A_1, B_2), (A_2, B_3), and (A_2, B_2) are used to check a CHSH inequality.
5. Alice and Bob perform error correction and privacy amplification to turn the sifted key into a shared secret key.

Exercise 4.3 Consider the Ekert protocol described above where instead of the state $|\Psi^-\rangle$, the state

$$|\Phi^+\rangle = \frac{1}{\sqrt{2}} \left(|00\rangle_{AB} + |11\rangle_{AB} \right) \tag{4.9}$$

is distributed, but the same measurement operators A_i, B_i are used.

1. Show that the measurement outcomes for the pairs (A_1, B_1) and (A_3, B_3), respectively, are completely correlated, i.e., show that whenever Alice obtains the result $+1$ (respectively, -1), Bob always obtains the result $+1$ (respectively, -1).
2. Show that the CHSH violation calculated from the pairs (A_1, B_3), (A_1, B_2), (A_2, B_3), (A_2, B_2) using the state $|\Phi^+\rangle$ is $S = 2\sqrt{2}$.
3. What is the value of S if Bob instead uses the measurement operators $B_2 = Z$ and $B_3 = X$?

Exercise 4.4 (Tsirelson's Bound) In this exercise, we want to show that

$$\mathrm{Tr}\big(\rho \left(A_1 \left(B_1 + B_2\right) + A_2 \left(B_1 - B_2\right)\right)\big) \leq 2\sqrt{2} \tag{4.10}$$

for any quantum state ρ and any set of self-adjoint operators A_1, A_2, B_1, B_2 with norm ≤ 1 that fulfil $[A_i, B_j] = 0$ for $i, j = 1, 2$. This is known as *Tsirelson's bound* [12]. Consider the operator

$$X = A_1 \left(B_1 + B_2\right) + A_2 \left(B_1 - B_2\right). \tag{4.11}$$

1. First, show that the statement $\|X\| \leq 2\sqrt{2}$ is equivalent to $\|X^2\| \leq 8$.
2. Show that $X^2 = 4 - [A_1, A_2][B_1, B_2]$. Use this result to compute an upper bound on $\|X^2\|$.
3. Use this approach to find a bound on the CHSH formula in (4.10) in case that Alice's observables commute.

4.1.2.2 Entanglement-Based Version of BB84

The entanglement-based version of the BB84 protocol[3] exploits the same ideas as the Ekert protocol, namely that a maximally entangled state leads to perfect correlations when the parties measure in the same basis. Suppose that Alice and Bob share the maximally entangled state

$$|\Phi^+\rangle_{AB} = \frac{1}{\sqrt{2}} \left(|00\rangle_{AB} + |11\rangle_{AB}\right). \tag{4.12}$$

Since this is a pure state, it cannot be entangled with anything else. In particular, this means that an eavesdropper does not have any information about the measurement results that are obtained from this state. Therefore, Alice and Bob's main goal is to distribute a sequence of m of these states, i.e.,

$$|\Phi^+\rangle_{AB}^{\otimes m} = |\Phi^+\rangle_{AB} \otimes \cdots \otimes |\Phi^+\rangle_{AB}, \tag{4.13}$$

and measure them in order to obtain a secret key that Eve cannot have any knowledge on (by the laws of quantum mechanics). However, in practice, they have to use an insecure quantum channel because of Eve's interactions with the quantum states and noise in general. Hence, they will not end up with the exact state given in (4.13), but rather with a mixed state ρ. Their task is then to correct the errors that were induced by transmitting the state over the quantum channel.

---

[3]This protocol is also referred to as the *modified Lo–Chau protocol*, since it exploits ideas introduced by Lo and Chau in [22].

To understand the details of this procedure, it is necessary to go into some detail about error correction codes, especially the so-called *Calderbank–Shor–Steane codes* (or simply CSS codes), which are quantum error correction codes that exploit the error-correcting properties of classical codes to detect and correct quantum errors. These codes are explained in more detail in Appendix B. We will also need these codes to show that the BB84 protocol is secure, but this is deferred to a later section. For now, it suffices to assume that Alice and Bob can use two classical error correction codes C_1 and C_2 to construct a CSS code that encodes m qubits into n qubits and that corrects arbitrary quantum errors (i.e., bit flips and phase errors and their combination) on up to t qubits. The protocol then works as follows:

Entanglement-Based BB84

1. Alice creates $2n$ qubit pairs in the state $|\Phi^+\rangle^{\otimes 2n}$ with $|\Phi^+\rangle$ as given in (4.12).
2. She randomly selects n of these qubits that will later be used to estimate the errors in the qubit pairs.
3. Alice selects a random classical bit string $b = (b_1, b_2, \ldots, b_{2n})$ of length $2n$. Whenever the bit b_i is 1, she applies a Hadamard transformation (2.28) to her half of the corresponding qubit pair.
4. She sends the other half of all qubit pairs to Bob.
5. Bob receives the qubits and publicly announces this fact.
6. Alice announces the string b and the positions of the check qubits.
7. Bob applies a Hadamard transformation to those qubits for which $b_i = 1$.
8. Alice and Bob measure the check qubits in the computational basis $\{|0\rangle, |1\rangle\}$ to estimate the error rate. If more than t errors occur, they abort the protocol.
9. If the number of errors is below t, Alice and Bob use the error correction codes C_1 and C_2 to correct the errors in the n remaining bits and obtain $|\Phi^+\rangle^{\otimes m}$.
10. They measure the state $|\Phi^+\rangle^{\otimes m}$ in the computational basis to obtain the shared secret key.

Note that applying the Hadamard transformation to the qubits before and after they are sent through the quantum channel has the same effect as preparing and measuring them in the Hadamard basis. We will show how exactly this protocol can be transformed into the original BB84 protocol as part of the security proof of the BB84 protocol, since this requires to discuss error correction codes in more detail.

4.1.3   From Prepare-and-Measure to Entanglement-Based Protocols

We have now discussed several prepare-and-measure protocols as well as entanglement-based protocols and already seen indications (for instance, the BB84 protocol) that these two types of protocols are connected. It can be useful to translate a prepare-and-measure protocol into an entanglement-based one since some security proof techniques require the latter type of protocol. In the entanglement-based version, Eve has more power because she controls both Alice's and Bob's part of the state, while in the prepare-and-measure version she cannot interact with Alice's state since it is never sent over a quantum channel. Therefore, security of the entanglement-based protocol automatically implies the security of the prepare-and-measure protocol.

Although transforming a prepare-and-measure protocol into an entanglement-based one is usually a non-trivial task since different protocols require making different assumptions, we can say that, in general, the connection is based on the following observations:

In a prepare-and-measure protocol, Alice's role is to choose a sequence of N symbols $x_1, x_2, \ldots, x_N$ that are realizations of a classical random variable X with probability distribution $p_X(x)$. She then encodes these symbols into a quantum state of the form

$$|\phi_{x_1}\rangle \otimes \cdots \otimes |\phi_{x_N}\rangle. \tag{4.14}$$

In all cases, it is crucial that non-orthogonal states are used for the encoding; otherwise, Eve is able to decode the sequence without introducing errors by measuring in the appropriate basis. Bob's role in the prepare-and-measure scheme is then to measure the quantum states that are sent by Alice.

This situation can equivalently be achieved by using entangled states in the following way: Instead of encoding a sequence of symbols into a quantum state that is sent to Bob, Alice prepares the bipartite entangled state

$$|\Phi\rangle_{AB} = \sum_x \sqrt{p_X(x)}|x\rangle_A \otimes |\phi_x\rangle_B, \tag{4.15}$$

where $\{|x\rangle_A\}$ is an orthonormal basis of the subsystem A, i.e., Alice's system. Alice then keeps the first half of the state in (4.15) and sends the other half to Bob. Alice measures the quantum system she kept with respect to the basis $\{|x\rangle\}_A$ to get the classical value X (which corresponds to the encoded bit, i.e., usually 0 or 1). It is easy to verify that the outcomes are distributed according to $p_X(x)$. The probability of obtaining an outcome y when measuring system A in the basis $\{|x\rangle_A\}$ while doing nothing on system B is given by

$$P(y) = {}_{AB}\langle\Phi|\left(|y\rangle_A\langle y|\otimes\mathbb{I}_B\right)|\Phi\rangle_{AB} \tag{4.16}$$

$$= \sum_{x,x'}\sqrt{p_X(x)}\sqrt{p_X(x')}\underbrace{{}_A\langle x|y\rangle_A}_{\equiv\delta_{x,y}}\underbrace{{}_A\langle y|x'\rangle_A}_{\equiv\delta_{y,x'}}{}_B\langle\phi_x|\phi_{x'}\rangle_B \tag{4.17}$$

$$= p_X(y). \tag{4.18}$$

The remaining quantum system, namely Bob's system, contains the correct encoding of X: If Alice obtains an outcome y after measuring her part of the state, then the state of Bob's system should be $|\phi_y\rangle$. The state of the system after Alice's measurement is given by

$$\frac{\left(|y\rangle_A\langle y|\otimes\mathbb{I}_B\right)|\Phi\rangle_{AB}}{P(y)} = \frac{1}{\sqrt{p_X(y)}}\sum_x\sqrt{p_X(x)}\,|y\rangle_A\underbrace{\langle y|x\rangle_A}_{\equiv\delta_{y,x}}\otimes|\phi_x\rangle_B \tag{4.19}$$

$$= \frac{1}{\sqrt{p_X(y)}}\sqrt{p_X(y)}\,|y\rangle_A\otimes|\phi_y\rangle_B \tag{4.20}$$

$$= |y\rangle_A\otimes|\phi_y\rangle_B, \tag{4.21}$$

which is exactly what we expect.

The fact that a prepare-and-measure scheme can be translated into an entanglement-based scheme does not mean that both schemes are equally practical or feasible with current technology. Rather, it implies that the security proof for the entanglement-based scheme automatically translates to a security proof for the prepare-and-measure protocol. This is very convenient, since the entanglement-based scheme only involves quantum states (and no quantum channels), which are usually easier to analyse. This is in fact the strategy that we will pursue to prove the security of the prepare-and-measure BB84 protocol in the next chapter: We will prove the security of the entanglement-based version of BB84 and then show that this formulation is equivalent to the prepare-and-measure scheme.

4.2 Classical Post-Processing

After the quantum transmission phase Alice and Bob hold a pair of bit strings that are, in general, partially correlated and partially secure. In the classical post-processing, they turn these bit strings into secure keys. The first step is the parameter estimation step: Alice and Bob have to estimate how many errors are there in their respective bit strings and with this the amount of information that was leaked to Eve during the quantum phase. If the error rate is above a certain threshold, they abort the protocol. Otherwise, they continue with the classical post-processing (see Fig. 4.5): First, they perform error correction, where their, in general, partly different bit strings are turned into identical ones. The second part is privacy amplification, where any residual information that Eve may have about the key is removed.

Fig. 4.5 Classical post-processing. If the output of the quantum phase of the protocol passes parameter estimation, classical post-processing turns the, in general, partially secret and partially correlated bit strings that Alice and Bob hold into secure keys. Error correction (EC) ensures that they hold identical bit strings, while privacy amplification (PA) minimizes Eve's knowledge on the key, hence making it secure

In all of these tasks, Alice and Bob use some part of their bit strings to achieve the respective goal. Here, it is important to find efficient methods such that only a small amount of bits has to be discarded for this purpose. In the remaining part of this section we will present some of the standard techniques that have been developed for these problems.

4.2.1 Parameter Estimation

After Alice and Bob have completed the quantum phase of the protocol, they want to make an estimate of the error rate in order to decide whether to abort the protocol because too much information was leaked to Eve or to proceed with transforming their bit strings into a secure key. For this purpose, they use a small sample of their bit strings to estimate a global property, namely the error rate, of those strings. The standard procedure here is that Alice sends a small sample of her string to Bob that he compares to his string. He then tells Alice the error rate he sees. If it is beyond a certain threshold, they abort the protocol, otherwise, they continue.

Although Alice and Bob only know the error rate for a small sample of their strings, this information enables them to make statements about the whole strings, which is due to the so-called *Chernoff–Hoeffding type bounds* [11, 19, 29]. These are statistical inequalities that state that given a random subset of data, a statistical property of the sample must be close to the statistical property of the whole data. This means that if Alice and Bob see an error rate of 7% in their sample, then the error rate of the whole strings is, with high probability, close to 7%.

We want to talk about these kinds of bounds and how to make use of them in the parameter estimation step in more detail. In our discussion we make use of a bound originally shown by Serfling [29], in a slightly different form presented in [1]:

Theorem 4.5 (Serfling's Inequality) *For a set of N random variables K_i with values $k_i \in \{0, 1\}$, where $i \in \{1, \ldots, N\}$, the average is defined as*

$$K = \frac{1}{N} \sum_{i=1}^{N} K_i. \tag{4.22}$$

Suppose we draw a sample (without replacement) of size n out of the set $\{K_i\}_i$ with values $x_j \in \{0, 1\}$, where $j \in \{1, \ldots, n\}$. Then its average is defined as

$$X = \frac{1}{n} \sum_{j=1}^{n} X_j. \tag{4.23}$$

Now let $k = N - n$ and $0 \leq \beta \leq 1$. Then

$$\Pr[X \geq K + \beta] \leq e^{-\frac{2\beta^2 nN}{k+1}}. \tag{4.24}$$

In brief, the above theorem states that the probability that the sample average X is bigger than the total average K is exponentially small in the sample size n. Therefore, we can make this probability smaller by enlarging the sample size.

How can we use this to find suitable bounds for parameter estimation? The quantity we are interested in is the probability that the error rate in the remaining n bits, denoted as Λ_n, is larger than the error rate that Alice and Bob observed in the k sampled bits, denoted as Λ_k. This probability is conditioned on the event that the error rate of the sampled bits, Λ_k, is below a certain threshold λ_{max}, since otherwise the protocol is simply aborted. Hence, the quantity we are interested in is

$$\Pr[\Lambda_n \geq \Lambda_k + \gamma | \Lambda_k \leq \lambda_{max}], \tag{4.25}$$

where γ is some small constant. The error rates are defined as follows. Alice and Bob's respective key strings split into a set of k bits that is used as the sample and a set of the remaining n bits. Therefore, their keys K_A and K_B can be written as

$$K_A = K_A^k K_A^n, \qquad K_B = K_B^k K_B^n. \tag{4.26}$$

Furthermore, $K_A^k \oplus K_B^k$ denotes the binary addition of the two sample bit strings. Note that the resulting string has 0s at those positions where Alice's and Bob's bits coincide and 1s at positions where their respective bits differ. The *Hamming weight* $|K_A^k \oplus K_B^k|$ of this string is the number of 1s, i.e., the number of errors in their sample bit strings (the same considerations can be done for K_A^n and K_B^n). The error rates are then defined as

$$\Lambda_n = \frac{1}{n} |K_A^n \oplus K_B^n|, \tag{4.27}$$

$$\Lambda_k = \frac{1}{k} |K_A^k \oplus K_B^k|. \tag{4.28}$$

Furthermore, using the notation $\nu = \frac{k}{N}$, the total error rate Λ (i.e., the one that includes both the sample bits and the remaining bits) can be written as

$$\Lambda = \frac{1}{N}|K_A \oplus K_B| = \nu\Lambda_k + (1 - \nu)\Lambda_n. \tag{4.29}$$

Since the k bits are randomly chosen by Alice and communicated to Bob all at once, they are sampled without replacement, which coincides with the requirement of the theorem.

We can now start to derive a bound on the quantity of interest. The first step is to conclude from Bayes' theorem (see Appendix A) that

$$\Pr[\Lambda_n \geq \Lambda_k + \gamma \,|\, \Lambda_k \leq \lambda_{\max}] \leq \frac{\Pr[\Lambda_n \geq \Lambda_k + \gamma]}{\Pr[\Lambda_k \leq \lambda_{\max}]}. \tag{4.30}$$

The term in the denominator on the right hand side is simply the probability that the protocol passes the check; hence, we will use the notation $p_{\text{pass}} = \Pr[\Lambda_k \leq \lambda_{\max}]$. The term in the nominator can be bounded as follows:

$$\Pr[\Lambda_n \geq \Lambda_k + \gamma] = \Pr[\nu\Lambda_n \geq \nu\Lambda_k + \nu\gamma] \tag{4.31}$$

$$= \Pr[\Lambda_n \geq \nu\Lambda_k + (1 - \nu)\Lambda_n + \nu\gamma] \tag{4.32}$$

$$= \Pr[\Lambda_n \geq \Lambda + \nu\gamma] \tag{4.33}$$

$$\leq e^{-\frac{2k^2 n\gamma^2}{(k+1)N}}, \tag{4.34}$$

where in the last step we have applied Theorem 4.5. We can now write (4.30) as

$$\Pr[\Lambda_n \geq \Lambda_k + \gamma \,|\, \Lambda_k \leq \lambda_{\max}] \leq \frac{e^{-\frac{2k^2 n\gamma^2}{(k+1)N}}}{p_{\text{pass}}}. \tag{4.35}$$

This inequality now states that the probability that the error rate on the remaining n bits is larger than the error rate on the k sample bits plus a small constant γ, given that the error rate is below the threshold $\lambda_{\max}$, is exponentially small in the sample size k. It quantifies the intuition that if the error rate in the sample is small, then the error rate in the remaining bits should be small too.

4.2.2 Error Correction

If Alice's and Bob's respective bit strings have passed the parameter estimation step, they proceed to correcting the errors. This step is also called *information reconciliation* because, in general, the objective is to turn two possibly different strings into two strings that are the same by possibly changing both of them.

However, in practice it is usually easier to consider the special case of error correction, where Bob changes his string in order to coincide with Alice's string.

In the last step, the parameter estimation step, Alice and Bob have estimated the error rate in their strings. Therefore, they now need to locate these errors in order to be able to correct them. A very simple error correction strategy is the following: Alice randomly chooses two bits from the sifted key and computes their XOR (i.e., exclusive OR) value. This is 0 if the two bits are the same and 1 if the two bits differ. She sends this value to Bob and tells him the positions of the corresponding bits. He compares this value with the XOR value he computes from his bits at the corresponding positions. If the values differ, Alice and Bob discard both bits. If the values are the same, they keep the first bit and discard the second. In this way, Eve does not gain any information on the actual key bit values. Although this strategy is easy to carry out, it is not very efficient. In case there are no errors in the bit strings, Alice and Bob still discard half of their bits, even more if there are errors in the strings.

Fortunately, there are more efficient error correction protocols. This is a well-studied research area that was investigated even before people came up with quantum cryptography. Therefore, there are explicit classical error correction codes that define how much communication is necessary for Bob to find and correct the errors. A variety of examples can be found in the books [23] and [18], for example.

Independent of the chosen error correction protocol, Alice and Bob have to make sure that the procedure was successful. Since Alice does not have access to Bob's system and Bob does not have access to Alice's system, neither of them can directly check if error correction was successful. One method they can use instead is applying a *two-universal hash function*.

Definition 4.6 Let $\mathcal{F}$ be a family of functions from an alphabet X to an alphabet Z and let p_F be a probability distribution on $\mathcal{F}$. The pair $(\mathcal{F}, p_F)$ is called two-universal if

$$\Pr_{f \in \mathcal{F}} [f(x) = f(x')] \leq \frac{1}{|Z|} \tag{4.36}$$

for any $x, x' \in X$ with $x \neq x'$ and f chosen randomly from $\mathcal{F}$ according to p_F.

Exercise 4.7 Show that the number $|\mathcal{F}|$ of functions in a family $\mathcal{F}$ of two-universal hash functions $f : X \to Z$ must, in general, be larger than the number $|Z|$ of keys they can generate.

For simplicity, we assume that p_F is the uniform distribution on $\mathcal{F}$. For alphabets $\{0, 1\}^n$ and $\{0, 1\}^l$ with $0 \leq l \leq n$, such a family of two-universal hash functions always exists. This is proven in [10] and [31], where explicit constructions for such families are given.

To check whether the error correction procedure was successful, Alice chooses at random a function from a family of two-universal hash functions and applies it to her bit string. She then sends both the function f_{EC} she chose and the output $f_{EC}(K_A)$ to Bob. Bob evaluates the function on his key and obtains $f_{EC}(K_B)$. He compares his result to Alice's output. If their hashes are equal, then with high probability their keys are the same. If their hashes differ, they abort the protocol.

We can show that the procedure that uses two-universal hash functions to check whether error correction was successful guarantees that the protocol is ϵ_{cor}-correct when using the right dimension of the output space of the hash functions. A protocol is ϵ_{cor}-correct if the probability that the two resulting keys of Alice and Bob differ is at most ϵ_{cor} (this is explained in more detail in Sect. 5.1). According to Definition 4.6, if the cardinality of the output space is $|\mathcal{Z}| = 2^{\lceil \log \frac{1}{\epsilon_{cor}} \rceil}$, we can show that the keys are equal except with probability ϵ_{cor}: First, note that it follows directly from the definition that

$$\Pr[f_{EC}(K_A) = f_{EC}(K_B)|K_A \neq K_B] \leq 2^{-\lceil \log \frac{1}{\epsilon_{cor}} \rceil} \leq \epsilon_{cor}. \tag{4.37}$$

This, together with Bayes' theorem (see Appendix A), implies that

$$\underbrace{\Pr[f_{EC}(K_A) = f_{EC}(K_B)|K_A \neq K_B]}_{\leq \epsilon_{cor}} \underbrace{\Pr[K_A \neq K_B]}_{\leq 1} \tag{4.38}$$

$$= \Pr[K_A \neq K_B|f_{EC}(K_A) = f_{EC}(K_B)] \underbrace{\Pr[f_{EC}(K_A) = f_{EC}(K_B)]}_{=1}, \tag{4.39}$$

where $\Pr[f_{EC}(K_A) = f_{EC}(K_B)] = 1$ follows from the fact that the protocol aborts if the hashes differ. In summary, the probability that the keys differ even though the hashes are the same is

$$\Pr[K_A \neq K_B|f_{EC}(K_A) = f_{EC}(K_B)] \leq \epsilon_{cor}. \tag{4.40}$$

Note that this checking procedure is independent of the error rate that Alice and Bob have observed in the parameter estimation step and also independent of the chosen error correction code. In the next section we will see that the privacy amplification procedure does not affect the correctness of the key, and show that if the checking procedure succeeds except with probability ϵ_{cor}, then the whole protocol is ϵ_{cor}-correct.

4.2.3   Privacy Amplification

The final task that Alice and Bob have to complete is to remove any knowledge that Eve has of the key after all the other steps in the protocol. This can be achieved by using the so-called *randomness extractors*. These are functions that take a source of randomness as input, for example, a string with a lower bound on its entropy, as well as a small uniformly random string (the *seed*), and output an almost uniformly random output that is longer than the seed. There are two additional requirements that we have: First, we want the seed and the output string to be independent of each other, since Alice has to communicate the seed publicly to Bob. This is guaranteed if we use a *strong* randomness extractor. Furthermore, we are not only interested in extracting randomness, but in extracting randomness with respect to a quantum adversary. Altogether, what we need is a *quantum-proof strong randomness extractor* [21].

To understand the definition of a quantum-proof strong randomness extractor we first need to clarify what systems and states are involved in this procedure. Alice's bit string is described by a classical random variable X, while Eve's information is represented by a quantum system E that is correlated with Alice's system. This situation can be described by the classical–quantum state[4]

$$\rho_{XE} = \sum_{x \in X} p_X(x)|x\rangle\langle x| \otimes \rho_E^x, \qquad (4.41)$$

where $\{|x\rangle\}$ is an orthonormal basis. This state is an element of $\mathcal{B}(\mathcal{H}_X \otimes \mathcal{H}_E)$. Furthermore, we need a state $\rho_Y \in \mathcal{B}(\mathcal{H}_Y)$ that describes the seed Y that is used to pick a function at random.

Finally, we have to specify the information that the adversary has on Alice's string. It is crucial that this information is limited, and the measure we use here is the *conditional min-entropy* we introduced in Definition 3.48. For a classical–quantum state as the one that describes Alice's and Eve's system, $H_{\min}(X|E)$ tells us how much uniform randomness $Z = f(X)$ we can extract from Alice's random variable X such that Z is independent of Eve's system E. If this quantity is bounded from below, we know that it is possible to extract some randomness. Z then forms the resulting secure key.

Definition 4.8 A (k, ϵ)-strong quantum-proof randomness extractor is a function $\mathsf{Ext} : \{0, 1\}^n \times \{0, 1\}^d \rightarrow \{0, 1\}^m$ if for all classical–quantum states ρ_{XE} with a classical random variable $X \in \{0, 1\}^n$ with min-entropy $H_{\min}(X|E) \geq k$ and a uniform random seed $Y \in \{0, 1\}^d$ we have

$$\frac{1}{2}\|\rho_{\mathsf{Ext}(X,Y)YE} - \frac{\mathbb{I}}{2^m} \otimes \rho_Y \otimes \rho_E\|_1 \leq \epsilon. \qquad (4.42)$$

It is clear that the state $\frac{\mathbb{I}}{2^m} \otimes \rho_Y \otimes \rho_E$ represents the ideal situation: The key, represented by the maximally mixed state $\frac{\mathbb{I}}{2^m}$, is uniformly random, i.e., each possible m-bit sequence has the same probability. Furthermore, it is independent of the seed Y and the state of Eve's system E, exactly as we required.

---

[4]Remember that we discussed these states in (2.141).

One example of a quantum-proof strong randomness extractor are two-universal hash functions as defined in Definition 4.6.[5] The strategy is similar to the application of two-universal hash functions described above: Alice and Bob publicly choose a function f_{PA} from a family of two-universal hash functions $\mathcal{F}$ at random (this is what they need the random seed Y for) and apply it to their key strings.[6] More precisely, let Alice's raw key after the error correction step be K_A, and then her resulting final key after the privacy amplification step is $f_{PA}(K_A)$. Since two-universal hash functions fulfil all the properties of a quantum-proof strong randomness extractor (see, for example, [27]), the resulting key is close to a perfect key, i.e., a uniformly random bit string that is independent of the seed and Eve's system. We can formulate an upper bound on the length of the resulting key in terms of the min-entropy by making use of the *Quantum Leftover Hash Lemma* [26, 30]:

Lemma 4.9 (Quantum Leftover Hash Lemma) *Let $\rho_{f_{PA}(K_A)YE}$ be the state after applying a random two-universal hash function f_{PA} to Alice's raw key K_A. Then for every $\epsilon' > 0$ it holds that*

$$D\left(\rho_{f_{PA}(K_A)YE}, \rho_U \otimes \rho_{YE}\right) \leq 2\epsilon' + \frac{1}{2}\sqrt{2^{l - H_{\min}^{\epsilon'}(K_A|E)}}, \tag{4.43}$$

where $\rho_U = \sum_{u \in \mathcal{Z}} \frac{1}{|\mathcal{Z}|}|u\rangle\langle u|$ is the maximally mixed state over the space of possible keys $\mathcal{Z}$.

Note that, in contrast to Definition 4.8, the above lemma uses the *smooth* min-entropy rather than the min-entropy because it gives a *tight* bound on the maximum amount of uniform randomness that can be extracted from K_A while being independent of E (see [20]). This gives rise to an additional smoothing parameter ϵ'. Together with the requirement that the protocol is ϵ_{sec}-secure, the above lemma gives an upper bound on the length l of the secure key:

$$l \leq H_{\min}^{\epsilon}(K_A|E) + 2 - \log \frac{1}{\epsilon_{\text{sec}} - 2\epsilon'}. \tag{4.44}$$

As a last point, since the correctness of the protocol is ensured by the error correction procedure, it is important that privacy amplification does not corrupt this correctness. If the keys K_A and K_B after the error correction step are the same (which is the case with probability at least ϵ_{cor}, as we just showed), then the outputs of the privacy amplification step are guaranteed to be the same. Let f_{PA} be the

---

[5] Another example is *Trevisan's extractor*, which is discussed in [15], for example.

[6] Since after the error correction step Alice and Bob's keys are identical with high probability, they can do the exact same steps on their bit strings to produce a pair of identical secure keys. Therefore, we only describe Alice's part in the following while keeping in mind that Bob performs the exact same actions on his raw key.

function used in the privacy amplification step. Then

$$\Pr[f_{\text{PA}}(K_A) \neq f_{\text{PA}}(K_B)] \leq \Pr[K_A \neq K_B] \leq \epsilon_{\text{cor}}. \tag{4.45}$$

Hence, for a suitable choice of the output space of the hash functions used in the error correction step, the whole protocol is ϵ_{cor}-correct.

This concludes our discussion of the classical post-processing phase. After Alice and Bob have performed the privacy amplification step and assuming that all steps have been carried out successfully, they hold two identical bit strings that are close to being uniformly random. Furthermore, Eve has almost no knowledge about the key. These are the requirements for a secure key that can be safely used in any cryptographic application.

References

1. Beaudry, N.J.: Assumptions in quantum cryptography. Ph.D. thesis, ETH Zurich (2014). https://doi.org/10.3929/ETHZ-A-010432410
2. Bechmann-Pasquinucci, H., Gisin, N.: Incoherent and coherent eavesdropping in the six-state protocol of quantum cryptography. Phys. Rev. A **59**(6), 4238–4248 (1999). https://doi.org/10.1103/physreva.59.4238
3. Bell, J.S.: On the Einstein Podolsky Rosen paradox. Phys. Phys. Fiz. **1**(3), 195–200 (1964). https://doi.org/10.1103/physicsphysiquefizika.1.195
4. Bennett, C.H.: Quantum cryptography using any two nonorthogonal states. Phys. Rev. Lett. **68**(21), 3121–3124 (1992). https://doi.org/10.1103/physrevlett.68.3121
5. Benatti, F., Fannes, M., Floreanini, R., Petritis, D. (eds.): Quantum Information, Computation and Cryptography. Springer, Berlin, Heidelberg (2010). https://doi.org/10.1007/978-3-642-11914-9
6. Branciard, C., Gisin, N., Kraus, B., Scarani, V.: Security of two quantum cryptography protocols using the same four qubit states. Phys. Rev. A **72**(3), 032301 (2005). https://doi.org/10.1103/physreva.72.032301
7. Bruß, D.: Optimal eavesdropping in quantum cryptography with six states. Phys. Rev. Lett. **81**(14), 3018–3021 (1998). https://doi.org/10.1103/physrevlett.81.3018
8. Bruß, D., Leuchs, G. (eds.): Lectures on Quantum Information. Wiley, New York (2006). https://doi.org/10.1002/9783527618637
9. Bruß, D., Meyer, T.: Quantum Cryptography. In: Quantum Information, Computation and Cryptography, pp. 277–308. Springer, Berlin, Heidelberg (2010). https://doi.org/10.1007/978-3-642-11914-9_9
10. Carter, J.L., Wegman, M.N.: Universal classes of hash functions. J. Comput. Syst. Sci. **18**(2), 143–154 (1979). https://doi.org/10.1016/0022-0000(79)90044-8
11. Chernoff, H.: A measure of asymptotic efficiency for tests of a hypothesis based on the sum of observations. Ann. Math. Stat. **23**(4), 493–507 (1952). https://doi.org/10.1214/aoms/1177729330
12. Cirel'son, B.S.: Quantum generalizations of Bell's inequality. Lett. Math. Phys. **4**(2), 93–100 (1980). https://doi.org/10.1007/bf00417500
13. Clauser, J.F., Horne, M.A., Shimony, A., Holt, R.A.: Proposed experiment to test local hidden-variable theories. Phys. Rev. Lett. **23**(15), 880–884 (1969). https://doi.org/10.1103/physrevlett.23.880
14. Curty, M., Lewenstein, M., Lütkenhaus, N.: Entanglement as a precondition for secure quantum key distribution. Phys. Rev. Lett. **92**(21) (2004). https://doi.org/10.1103/physrevlett.92.217903

15. De, A., Portmann, C., Vidick, T., Renner, R.: Trevisan's extractor in the presence of quantum side information. SIAM J. Comput. **41**(4), 915–940 (2012). https://doi.org/10.1137/100813683
16. Ekert, A.K.: Quantum cryptography based on Bell's theorem. Phys. Rev. Lett. **67**(6), 661–663 (1991). https://doi.org/10.1103/physrevlett.67.661
17. Gisin, N., Ribordy, G., Tittel, W., Zbinden, H.: Quantum cryptography. Rev. Mod. Phys. **74**(1), 145–195 (2002). https://doi.org/10.1103/revmodphys.74.145
18. Hamming, R.W.: Coding and Information Theory. Prentice-Hall, Englewood Cliffs, NJ (1986)
19. Hoeffding, W.: Probability inequalities for sums of bounded random variables. J. Am. Stat. Assoc. **58**(301), 13–30 (1963). https://doi.org/10.1080/01621459.1963.10500830
20. Konig, R., Renner, R., Schaffner, C.: The operational meaning of min- and max-entropy. IEEE Trans. Inform. Theory **55**(9), 4337–4347 (2009). https://doi.org/10.1109/tit.2009.2025545
21. König, R., Renner, R.: Sampling of min-entropy relative to quantum knowledge. IEEE Trans. Inform. Theory **57**(7), 4760–4787 (2011). https://doi.org/10.1109/tit.2011.2146730
22. Lo, H.K., Chau, H.F.: Unconditional security of quantum key distribution over arbitrarily long distances. Science **283**(5410), 2050–2056 (1999). https://doi.org/10.1126/science.283.5410.2050
23. MacWilliams, F.J.: The Theory of Error Correcting Codes. North-Holland Publication, Amsterdam, New York (1977)
24. Nielsen, M.A., Chuang, I.L.: Quantum Computation and Quantum Information. Cambridge University Press, Cambridge (2000)
25. Pirandola, S., Andersen, U.L., Banchi, L., Berta, M., Bunandar, D., Colbeck, R., Englund, D., Gehring, T., Lupo, C., Ottaviani, C., Pereira, J., Razavi, M., Shaari, J.S., Tomamichel, M., Usenko, V.C., Vallone, G., Villoresi, P., Wallden, P.: Advances in quantum cryptography. Adv. Opt. Photonics **12**(4), 1012–1236 (2019). https://doi.org/10.1364/AOP.361502
26. Renner, R.: Security of quantum key distribution. Ph.D. thesis, ETH Zurich (2005). https://doi.org/10.3929/ethz-a-005115027
27. Renner, R., König, R.: Universally Composable Privacy Amplification Against Quantum Adversaries. In: Theory of Cryptography, pp. 407–425. Springer, Berlin, Heidelberg (2005). https://doi.org/10.1007/978-3-540-30576-7_22
28. Scarani, V., Acín, A., Ribordy, G., Gisin, N.: Quantum cryptography protocols robust against photon number splitting attacks for weak laser pulse implementations. Phys. Rev. Lett. **92**(5), 057901 (2004). https://doi.org/10.1103/physrevlett.92.057901
29. Serfling, R.J.: Probability inequalities for the sum in sampling without replacement. Ann. Stat. **2**(1), 39–48 (1974). https://doi.org/10.1214/aos/1176342611
30. Tomamichel, M., Schaffner, C., Smith, A., Renner, R.: Leftover hashing against quantum side information. IEEE Trans. Inform. Theory **57**(8), 5524–5535 (2011). https://doi.org/10.1109/tit.2011.2158473
31. Wegman, M.N., Carter, J.L.: New hash functions and their use in authentication and set equality. J. Comput. Syst. Sci. **22**(3), 265–279 (1981). https://doi.org/10.1016/0022-0000(81)90033-7

Security Analysis

5

Abstract

Proving the security of a quantum key distribution protocol is crucial with regard to using the protocol in any practical application. Without a security proof one cannot be sure that the created key can safely be used for communication tasks. The security analysis of a quantum key distribution protocol involves several steps: the final goal is to make a security claim about the protocol. For this purpose, it is necessary to first give a precise definition of what we mean by "security". To arrive at the security claim, we first clarify the assumptions that we make for the protocol in question. There are some assumptions we have to make independent of the protocol (e.g., that quantum mechanics is correct), but others are very specific, such as a claim about the efficiencies of the included detectors. Furthermore, we need to clarify what kind of attacks Eve can perform. A general security proof should take into account the most powerful attacks that she can do. Starting from these assumptions we can formulate the security proof. This proof can take different forms, depending on the assumptions we make, the protocol we use, and what kind of techniques are applicable.

5.1 Definition of Security

Before we can formulate a security proof for an actual protocol based on assumptions we make (see Fig. 5.1), we need to define what we actually mean by security. Intuitively, for a QKD protocol to be secure we need to ensure that the information that Eve has on the resulting key is negligible. One measure of Eve's information is the mutual information between the key K that Alice and Bob share after the protocol and a random variable W that describes the outcome of a measurement that Eve applies to her system after the protocol. A possible security criterion is

© The Author(s), under exclusive license to Springer Nature Switzerland AG 2021 117
R. Wolf, *Quantum Key Distribution*, Lecture Notes in Physics 988,
https://doi.org/10.1007/978-3-030-73991-1_5

Fig. 5.1 Security proof of a
QKD protocol. Any security
proof of a QKD protocol is
based on the assumptions we
make about the systems,
states, and measurements that
are involved. The goal of the
proof is to make a claim about
the security of the protocol

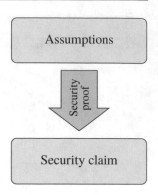

then the maximum of this quantity over all possible measurements with output W
that Eve can apply:

$$\max_{W} I(K : W) \leq \epsilon. \tag{5.1}$$

At first sight, this definition looks reasonable: since the mutual information between
two random variables quantifies the correlation between the two systems, it seems
that it captures the intuitive meaning of security. Although this definition looks
reasonable at first sight, it does not guarantee universal security in the way we
demand it. More precisely, this criterion does not guarantee that the key can safely
be used in any application. This is because Eve can, in principle, wait with her
measurement until she learns parts of the key. A more detailed discussion of this
concern with respect to existing security definitions can be found in [3] and [28].

To illustrate the kind of problems that can appear, consider the following
scenario: Assume that Alice has a key $K = (K_1, K_2, \ldots K_n)$ of length n that
she uses to encode an n-bit message $M = (M_1, M_2, \ldots, M_n)$ by one-time pad
encryption (see Fig. 1.1). Recall that the resulting ciphertext $C = (C_1, C_2, \ldots, C_n)$
that Alice sends to Bob is the bitwise XOR of K and M, i.e.,

$$C_i = K_i \oplus M_i, \tag{5.2}$$

where $\oplus$ is the binary addition. Suppose Eve is interested in the n-th bit of the
message, M_n, and she already knows the first $n-1$ bits $M_1, M_2, \ldots, M_{n-1}$.[1] Using
this knowledge, she can easily determine the first $n-1$ bits of the key K, simply by
doing binary addition:

$$K_i = M_i \oplus C_i. \tag{5.3}$$

---

[1] This situation is not as artificial as it may seem at first sight. The first bits can be some redundant
header information before the actual message. In fact, correlations between parts of the key are the
reason why many encrypted messages have been decrypted in history. It played an important role,
for instance, in the decryption of the Enigma during the Second World War.

In order to guarantee the security of the n-th bit of the message, it is crucial that Eve still does not have any knowledge of the n-th key bit K_n, even though she already knows the first $n - 1$ bits, i.e., the individual bits of the key are completely uncorrelated. This requirement, however, is not covered by the definition of security in (5.1). For an arbitrary ϵ and an n that depends on ϵ, it is possible to construct scenarios where the security criterion (5.1) is fulfilled, but Eve can determine the n-th bit of the key, K_n, with certainty once she knows the first $n - 1$ bits of the key (see [17] for more details).

Therefore, we need a stronger definition of security that ensures the safe use of the key in all applications. In particular, we need to make sure that the key is still secure when it is used as a component within a bigger system. This is called *universal security*: a key is said to be universally secure if it is secure in any arbitrary context. This especially implies that any bit of the key K remains secret even if some other part of K is revealed.

A Universal Security Definition

The idea behind universal security definitions is usually to characterize the real cryptographic protocol by its distance to an ideal one (which is perfectly secure by definition). We can divide the criterion for a secure QKD protocol into two parts: the *correctness* of the key and the *secrecy*.

Defining correctness is straightforward: we simply require that the probability that Alice and Bob do not abort the protocol *and* the generated keys are different is small. However, there is one point where we have to be careful: for the definition to also be applicable to the case where the protocol aborts, we have to add a new symbol to the key space S (i.e., the space that is made of all possible sequences the key bit string can take): If the protocol aborts, we write $K_A = \perp$ and $K_B = \perp$ to denote that Alice and Bob know that the protocol aborted.

Definition 5.1 (ϵ-Correctness) Let K_A and K_B be the random variables that describe Alice's and Bob's respective key bit strings (i.e., random variables over the alphabet $S \cup \perp$) at the end of the protocol. The protocol is said to be ϵ-correct if

$$\Pr[K_A \neq K_B] \leq \epsilon. \tag{5.4}$$

For the secrecy part, we compare the real key to a *perfect key*, where perfect means that the key is uniformly distributed and independent of the adversary's information: consider Alice's key K_A,[2] which is distributed according to the probability distribution p_{K_A}. Furthermore, let $\rho_E^{k_A}$ be the state of Eve's system given that the key takes the value k_A, for any element $k_A \in S$, where S is the key space. The joint state of the (classical) key K_A and Eve's system can be written as the

---

[2]Note that the same considerations can be done for Bob's key since the keys are the same except with probability ϵ_{cor}.

classical–quantum state

$$\rho_{K_A E} = \sum_{k_A \in S} p_{K_A}(k_A) |k_A\rangle\langle k_A| \otimes \rho_E^{k_A}, \tag{5.5}$$

where $\{|k_A\rangle\}_{k_A \in S}$ is an orthonormal basis of some Hilbert space $\mathcal{H}_{K_A}$. It is now helpful to consider the distance between the state of the real scenario and the state of the ideal scenario individually for the two possible cases (abort and pass): first, notice that Eve's state is the same in both cases since aborting the protocol does not affect her system. If the protocol aborts, then Alice's state is trivial and since Eve's state is the same for both the real and ideal protocols, the distance is zero. In the case where the protocol does not abort, the ideal scenario is the one where Eve has no information on Alice's system and Alice's state is uniformly random:

$$\tilde{\rho}_{K_A E}^{\text{pass}} = \rho_U \otimes \rho_E, \tag{5.6}$$

where $\rho_U = \sum_{u \in S} \frac{1}{|S|} |u\rangle\langle u|$ is the fully mixed state on $\mathcal{H}_U$. Furthermore, let $p^\perp$ be the probability that the protocol aborts. We can now use the triangle inequality of the trace distance to estimate the distance between the real state $\rho_{K_A E}$ and the ideal state $\tilde{\rho}_{K_A E}$:

$$\|\rho_{K_A E} - \tilde{\rho}_{K_A E}\|_1 \leq p^\perp \cdot 0 + (1 - p^\perp) \|\rho_{K_A E}^{\text{pass}} - \rho_U \otimes \rho_E\|_1. \tag{5.7}$$

The secrecy of the key K_A with respect to the adversary E is then defined as follows:

Definition 5.2 (ϵ-Secrecy) A key K_A is said to be ϵ-secret if for any state $\rho_{K_A E}^{\text{pass}}$ the state of the composite system of Alice and Eve after a QKD protocol, conditioned on the event that the protocol does not abort, satisfies

$$(1 - p^\perp)\frac{1}{2}\|\rho_{K_A E}^{\text{pass}} - \rho_U \otimes \rho_E\|_1 \leq \epsilon, \tag{5.8}$$

where $\rho_U = \sum_{u \in S} \frac{1}{|S|} |u\rangle\langle u|$.

As we argued above, the security of a QKD protocol comprises both correctness and secrecy.[3] Combining these two criteria then yields the following definition of security:

Definition 5.3 (ϵ-Security) Let $\rho_{K_A K_B E}^{\text{pass}}$ be the state of the system shared between Alice, Bob, and Eve after a QKD protocol, conditioned on the event that the protocol

---

[3]Throughout the literature, there is a variety of different definitions of security. Sometimes it is defined only as secrecy, and sometimes it is a combination of correctness, secrecy, and robustness. In these notes we choose to define security as correctness and secrecy, while robustness is given as an additional criterion.

does not abort, i.e.,

$$\rho^{\text{pass}}_{K_A K_B E} = \frac{1}{1 - p^{\perp}} \sum_{k_A, k_B \in S} p_{K_A, K_B}(k_A, k_B) |k_A k_B\rangle\langle k_A k_B| \otimes \rho_E^{k_A, k_B}, \tag{5.9}$$

where $p^{\perp}$ is the probability that the protocol aborts. Then the protocol is said to be ϵ-secure under any attack performed by Eve if the state satisfies

$$(1 - p^{\perp}) \cdot \frac{1}{2} ||\rho^{\text{pass}}_{K_A K_B E} - \rho_{UU} \otimes \rho_E||_1 \leq \epsilon, \tag{5.10}$$

where $\rho_{UU} = \sum_{u \in S} \frac{1}{|S|} |u\rangle\langle u| \otimes |u\rangle\langle u|$ for some family $\{|u\rangle\}_{u \in S}$ of orthonormal vectors that represent the values of the key space S.

This definition ensures that the key is universally secure in the following way: (5.10) guarantees that the actual situation, which is represented by the state $\rho^{\text{pass}}_{K_A E}$, is ϵ-close to an ideal situation. The ideal situation is thereby described by the state $\rho_{UU} \otimes \rho_E$, where ρ_U represents the perfect key that is independent of the state ρ_E of Eve's system. In particular, the perfect key U is uniformly distributed, which implies that each sequence is equally probable. To illustrate this definition and the role of the parameter ϵ in the above definition, we compare the perfect key and the real key in Fig. 5.2.

Some further remarks on the definition of security: note that we do not need to define security conditioned on not aborting by the same argument we have made above when defining ϵ-secrecy. Also, note that the fact that the trace distance cannot increase when applying a quantum operation (as we have seen in Theorem 2.56) ensures that an ϵ-secure key will remain secure for any possible evolution. Since the

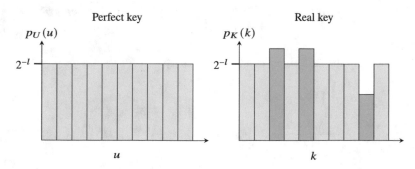

Fig. 5.2 ϵ-security. A perfect key is one where every possible key sequence is equally probable: for a key of length l, there are 2^l possibilities; hence, the probability that a specific sequence u occurs is 2^{-l}. A real key that is ϵ-close to a perfect key contains some sequences whose probability deviates from the uniform distribution. The parameter ϵ describes how much weight has to be moved to transform the real distribution to the uniform distribution

above definition of security does not make any assumptions about the state that is shared by Alice, Bob, and Eve, it applies to any attack that Eve can perform.

There is a practical relationship between correctness, secrecy, and the above definition of security that makes it easy to define security directly from the definitions of correctness and secrecy. It is summarized by the following theorem:

Theorem 5.4 *If a protocol is ϵ_{cor}-correct and ϵ_{sec}-secret, then it is ϵ-secure with $\epsilon = \epsilon_{\text{cor}} + \epsilon_{\text{sec}}$.*

Proof For simplicity, we use the following notation for the trace distance:

$$D(\rho, \sigma) = \frac{1}{2} \|\rho - \sigma\|_1. \tag{5.11}$$

To show that the protocol is $\epsilon_{\text{cor}} + \epsilon_{\text{sec}}$-secure, we have to show that the following inequality holds:

$$(1 - p^{\perp}) D\left(\rho_{K_A K_B E}^{\text{pass}}, \rho_{UU} \otimes \rho_E\right) \leq \epsilon_{\text{cor}} + \epsilon_{\text{sec}}. \tag{5.12}$$

For this purpose, we introduce the state σ_{ABE}, which is obtained from the state $\rho_{K_A K_B E}^{\text{pass}}$ by throwing away Bob's system B and replacing it with a copy of Alice's system A; hence, it is of the form

$$\sigma_{ABE} \equiv \frac{1}{1 - p^{\perp}} \sum_{k_A, k_B \in S} p_{K_A K_B}(k_A, k_B) |k_A k_A\rangle \langle k_A k_A| \otimes \rho_E^{k_A, k_B}. \tag{5.13}$$

We then use the fact that the trace distance fulfils the triangle inequality; hence,

$$D\left(\rho_{K_A K_B E}^{\text{pass}}, \rho_{UU} \otimes \rho_E\right) \leq D\left(\rho_{K_A K_B E}^{\text{pass}}, \sigma_{ABE}\right) + D\left(\sigma_{ABE}, \rho_{UU} \otimes \rho_E\right). \tag{5.14}$$

The first term in the sum above can be bounded by

$$D\left(\rho_{K_A K_B E}^{\text{pass}}, \sigma_{ABE}\right) \tag{5.15}$$

$$\leq \sum_{k_A, k_B \in S} \frac{p_{K_A K_B}(k_A, k_B)}{1 - p^{\perp}} D\left(|k_A k_B\rangle \langle k_A k_B| \otimes \rho_E^{k_A, k_B}, |k_A k_A\rangle \langle k_A k_A| \otimes \rho_E^{k_A, k_B}\right) \tag{5.16}$$

$$= \sum_{k_A \neq k_B} \frac{p_{K_A K_B}(k_A, k_B)}{1 - p^{\perp}} \tag{5.17}$$

$$= \frac{1}{1 - p^{\perp}} \Pr[K_A \neq K_B]. \tag{5.18}$$

To arrive at (5.17), we have used that two cases can occur: if $k_A = k_B$, then the trace distance in the corresponding term in (5.16) is equal to zero. If $k_A \neq k_B$, then the trace distance in (5.16) is equal to one. For the second term in (5.14) note that in both states σ_{ABE} and $\rho_{UU} \otimes \rho_E$ system B is a copy of system A; hence, it does not have an effect on the trace distance. On the other hand, system B is the only system that is different in the states $\rho^{\mathrm{pass}}_{K_A K_B E}$ and σ_{ABE}; therefore,

$$\mathrm{Tr}_B \left(\rho^{\mathrm{pass}}_{K_A K_B E} \right) = \mathrm{Tr}_B \left(\sigma_{ABE} \right). \tag{5.19}$$

Combining these two observations yields the following identities:

$$D\left(\sigma_{ABE}, \rho_{UU} \otimes \rho_E\right) = D\left(\sigma_{AE}, \rho_U \otimes \rho_E\right) = D\left(\rho^{\mathrm{pass}}_{K_A E}, \rho_U \otimes \rho_E \right). \tag{5.20}$$

We can now insert the above results into (5.14) and obtain

$$(1 - p^\perp)D\left(\rho^{\mathrm{pass}}_{K_A K_B E}, \rho_{UU} \otimes \rho_E \right) \leq \Pr[K_A \neq K_B] + (1 - p^\perp)D\left(\rho^{\mathrm{pass}}_{K_A E}, \rho_U \otimes \rho_E \right) \tag{5.21}$$

$$\leq \epsilon_{\mathrm{cor}} + \epsilon_{\mathrm{sec}}, \tag{5.22}$$

which proves the statement. □

Robustness

As we have argued above, Definition 5.3 now matches our requirements for a definition of security for a quantum key distribution protocol. However, although it guarantees that the keys that are generated by the protocol are always secure, it does not guarantee that the protocol is useful. This becomes clear when we consider the protocol that always aborts, which is secure in the sense of the above definition. Although this situation fits to the notion of security (because it never outputs an insecure key), we do not regard it as a useful protocol.

Therefore, the missing ingredient here is the notion of *robustness*, which guarantees that, in the absence of an eavesdropper, the protocol succeeds with high enough probability:

Definition 5.5 (ϵ-Robustness) A QKD protocol is said to be ϵ-robust if, in the absence of an eavesdropper, the probability that the protocol aborts is

$$p^\perp_{\text{no Eve}} = \epsilon. \tag{5.23}$$

Composability

One aspect that we have not mentioned yet, but which is important for practical applications, is *composability*. It is important that a protocol can be composed with another protocol in a way that the security of the protocols is preserved. Here, we must distinguish between two scenarios: *Sequential* composition is where the output of the first protocol is used as input for the second protocol, for example, when we compose a one-time pad with a QKD protocol. In the second case, *parallel* composition, two protocols are executed simultaneously and are combined to be treated as one protocol.

Sequential composition can be proved by using the triangle inequality for the trace distance: if one protocol is ϵ_1-secure and the second is ϵ_2-secure, then the whole protocol is $\epsilon_1 + \epsilon_2$-secure. The second scenario, parallel composition, can be proved in a similar way. We do not go into further details here but refer to [20, 21] and [24].

5.2   Assumptions

To prove the security of a protocol, it is important to be clear about the assumptions that one makes. Most of the assumptions account for the properties and imperfections of implementations of the protocol. So far, we have usually assumed *idealized* protocols,[4] where everything works exactly as described. However, there are some foundational assumptions that apply to almost all quantum cryptographic settings:

1. **Quantum theory is correct.** If the underlying theory is false, there is no way to create a scheme that is secure in actual applications. Therefore, we assume that quantum mechanics makes accurate predictions about measurement outcomes.
2. **Quantum theory is complete.** It is important that the theory we use explains all phenomena that can appear. This ensures that an adversary cannot get more information on Alice's and Bob's respective keys than what is possible with quantum mechanics. Interestingly, the correctness of quantum mechanics together with the requirement that *free randomness exists*[5] directly implies that the theory is also complete, which was shown in [11].[6]
3. **Authentic communication is possible.** This means that Alice and Bob can authenticate themselves, i.e., Eve cannot replace one of them without being detected.

Apart from these fundamental assumptions about the underlying physical theory, there are many assumptions one can make that apply to the implementation of the protocol. Any deviation of the implementation of a protocol from the theoretic model under which the security was proved may yield a leakage of information to Eve, which may cause the entire protocol to be insecure, since Eve might attack

---

[4]One exception is the SARG04 protocol in Sect. 4.1.1 that accounts for a weakness of the implementation of single-photon sources via weak laser pulses. Here, in contrast to the other protocols we have described, we did not assume that Alice and Bob have access to a perfect single-photon source.

[5]This requirement simply means that measurement choices (for instance, the choice of basis in the BB84 protocol) can be made independently of the measurement device itself.

[6]A more accessible explanation of this connection can be found in [27].

the implementation in a way that is not accounted for in the model. Hence, it is not sufficient to prove security for an idealized model: it is crucial that the model accurately describes the practical implementation. For this purpose, it is important to be clear about all assumptions one makes on the protocol, especially with respect to the implementation. An exhaustive treatment of assumptions in quantum cryptography can be found in [1] and [29]. Here, we will only give a short overview of the different ways in which ideal protocols can differ from actual implementations:

1. **Isolation of Alice's and Bob's labs.** In the idealized protocols we usually assume that Alice's and Bob's respective labs are isolated, in particular that Eve cannot tamper with the sources and measurement devices, while on the other hand, we assume that Eve has access to the quantum channel. However, since Alice and Bob use the quantum channel to send and receive states they need to have some sort of interface with the channel. Here, it might be possible that Eve can exploit this interface to probe Alice's and Bob's devices and gain additional information on the prepared states and the measurement outcomes in this way.

2. **Preparation of states.** In the protocols we have described in the previous chapter we always assume that Alice is able to prepare exact states. However, in practice it is not possible to prepare states with arbitrary precision; hence, the actual states can differ from the specific states used in the theoretical model.

3. **Measurement devices.** Similar to the state sources, the measurement devices usually do not act exactly how it is specified by the POVM elements. It is also possible that they not only react to the states specified by the protocol, but also to states outside the Hilbert space they are supposed to measure in. Furthermore, the measurement device can output results that are not part of the protocol, for example if there are losses in the device such that there is no result even though a quantum state was prepared and sent via the quantum channel.

   Another important aspect of measurement devices in practical implementations that is not accounted for in the theoretical model is that they have to be calibrated before they can be used in the actual protocol, which may give additional information about the devices to Eve.

4. **Timing.** We have often used that Alice and Bob compare their bits at a specific position in the string, especially in classical post-processing. To be able to talk about their strings in this way, they need to agree on the timing of signals so that sent states are associated with the correct measurement outcomes. Furthermore, measurement devices usually have a dead time after a measurement process where they do not respond to an incoming signal.

5. **Classical post-processing.** An important aspect of the classical post-processing part of a protocol is to estimate how much information was leaked to Eve. If this estimation does not include the differences between the ideal model and the actual implementation, it is possible that more information is leaked to Eve than the security proof accounts for.

A way to avoid having to deal with all the intricate assumptions about labs and devices is provided by *device-independent* QKD. Here, (almost) no assumptions about the devices used by Alice and Bob are made; hence, the security proof also applies in the case of corrupted or malfunctional devices. Instead, the security of the protocol relies on the violation of a Bell inequality, similar to the Ekert protocol in Sect. 4.1.2, where it is crucial that the CHSH inequality is violated. We explain the concept of device-independent QKD in more detail in the next chapter.

5.3   Eavesdropping Strategies

Before we explain how the security of a protocol can be proven, we discuss the kind of attacks that an eavesdropper can do. We have already seen an example of an attack that Eve can do, namely the *intercept-and-resend* strategy presented in the previous chapter. Let us now consider a slightly more general (but not the most general) attack at the BB84 protocol that shows the trade-off between the disturbance of the system that Eve introduces by interacting with the states that Alice sends and the information gain she achieves by this interaction. Recall that Alice always sends one of the four states $|\psi_{00}\rangle, |\psi_{10}\rangle, |\psi_{01}\rangle, |\psi_{11}\rangle$ defined in (4.2)–(4.5), which are not all mutually orthogonal. To get information about a state, Eve attaches an ancilla system in some predefined state $|E\rangle$[7] to the qubit sent by Alice and applies a unitary operation U to the composite system.

First, we look at how much information Eve can gain from a measurement that does not disturb at all the states sent by Alice, which means that Eve can perform this attack without Alice and Bob being able to detect it. Consider the action of this attack on two states that are non-orthogonal, for instance, the states $|\psi_{01}\rangle$ and $|\psi_{10}\rangle$:

$$U|\psi_{01}\rangle|E\rangle = |\psi_{01}\rangle|E_{\psi_{01}}\rangle \tag{5.24}$$

$$U|\psi_{10}\rangle|E\rangle = |\psi_{10}\rangle|E_{\psi_{10}}\rangle, \tag{5.25}$$

where $|E_{\psi_{01}}\rangle$ denotes the state of the ancilla system after the unitary operation in case Alice's state is $|\psi_{01}\rangle$ (and analogously for the other case). Unitary operations preserve the scalar product. We can now compare the scalar products of the left hand sides and the right hand sides of (5.24) and (5.25) to get some information on the state of the ancilla system after the unitary operation:

$$\langle\psi_{01}|\psi_{10}\rangle\langle E|E\rangle = \langle\psi_{01}|\psi_{10}\rangle\langle E_{\psi_{01}}|E_{\psi_{10}}\rangle. \tag{5.26}$$

Since $\langle E|E\rangle = 1$, the above equation implies that $\langle E_{\psi_{01}}|E_{\psi_{10}}\rangle = 1$ and therefore $|E_{\psi_{01}}\rangle$ and $|E_{\psi_{10}}\rangle$ are identical. Hence, the state of the ancilla system does not carry

---

[7]Note that the initial quantum state of the ancilla system is not important here. The symbol E is simply a place holder.

any information about Alice's quantum state, which shows that Eve cannot gain information by using an attack that does not disturb Alice's and Bob's states at all.

Therefore, Eve has to interact with Alice's state in a way that introduces a disturbance to the state, i.e., the unitary operation she applies has the following effect:

$$U|\psi_{01}\rangle|E\rangle = |\psi'_{01}\rangle|E_{\psi_{01}}\rangle \tag{5.27}$$

$$U|\psi_{10}\rangle|E\rangle = |\psi'_{10}\rangle|E_{\psi_{10}}\rangle. \tag{5.28}$$

In this case, the scalar product is

$$\langle\psi_{01}|\psi_{10}\rangle\langle E|E\rangle = \langle\psi'_{01}|\psi'_{10}\rangle\langle E_{\psi_{01}}|E_{\psi_{10}}\rangle. \tag{5.29}$$

In order to gain information about the states sent by Alice, Eve needs the states of the ancilla system to be distinguishable. The smaller the scalar product $\langle E_{\psi_{01}}|E_{\psi_{10}}\rangle$ is, the more distinguishable are the two states. However, if the scalar product of the ancilla states decreases, the scalar product of Alice's states has to increase in order to fulfil (5.29). In other words, the more information Eve gains, the greater the disturbance she introduces to Alice's states. This in turn means that the attack can be more easily detected by Alice and Bob.

5.3.1 Classification of Attacks

In general, the attacks that Eve can use to get information about the key can be divided into three different classes: in increasing order of power given to Eve, the attacks are *individual*, *collective*, and *coherent* attacks. Individual attacks are the simplest ones with only little power given to Eve, while coherent ones are the most powerful attacks where we assume that Eve has unlimited power and resources and is only limited by the laws of nature. Individual and collective attacks are usually considered in order to simplify the security analysis, although it is necessary to also consider coherent attacks in order to prove security for a QKD protocol. However, it might still be useful to study a protocol with respect to the two less powerful attacks to check if it is secure at all or if there is a simple attack strategy for Eve that lets her gain information without introducing a disturbance to the quantum states sent by Alice.

The different types of attacks can be described by the way Eve interacts with the quantum states sent by Alice and how Eve processes the information she gets in this way, which is summarized in Table 5.1. The procedure of extracting information from a quantum system can, in general, be described as follows: Eve attaches an ancilla system in the predefined state $|E\rangle_E\langle E|$ to the state ρ_A that Alice sends. Then she performs a unitary operation U on the composite system, which leaves the state

Table 5.1 Summary of Eve's attacks. Eve can attack the states $\rho_A^1, \ldots, \rho_A^n$ that Alice sends in three different ways: Individually, where the same unitary and the same POVM are applied to each state, collectively, where the same unitary is applied to each state but the measurement is given by a global POVM, or coherently, where a global ancilla system is attached to the tensor product of Alice's states. $\mathcal{M}^1$ denotes a POVM that describes the measurement of a single system, and $\mathcal{M}^n$ denotes the global POVM

	Ancilla state	Prob. dist.
Individual	$\rho_E^i = \mathrm{Tr}_A\left(U^\dagger\left(\rho_A^i \otimes \lvert E\rangle_E\langle E\rvert\right)U\right)$	$P_{\mathcal{M}^1}^{\rho_E^1} \ldots P_{\mathcal{M}^1}^{\rho_E^n}$
Collective	$\rho_E^i = \mathrm{Tr}_A\left(U^\dagger\left(\rho_A^i \otimes \lvert E\rangle_E\langle E\rvert\right)U\right)$	$P_{\mathcal{M}^n}^{\rho_E^1 \otimes \cdots \otimes \rho_E^n}$
Coherent	$\rho_E = \mathrm{Tr}_A\left(U_G^\dagger\left((\rho_A^1 \otimes \cdots \otimes \rho_A^n) \otimes \lvert E\rangle_E\langle E\rvert\right)U_G\right)$	$P_{\mathcal{M}^n}^{\rho_E}$

of the ancilla system in the form

$$\rho_E = \mathrm{Tr}_A\left(U^\dagger \rho_A \otimes \lvert E\rangle_E\langle E\rvert U\right). \tag{5.30}$$

Afterwards, Eve measures the ancilla system, which is given by a POVM $\mathcal{M} = \{M_j\}$, where outcome j is obtained with probability $\mathrm{Tr}\left(M_j\rho\right)$ when measuring a system in the state ρ.

We now consider the scenario where Alice sends n quantum states $\rho_A^1, \rho_A^2, \ldots,$ ρ_A^n to Bob. In case Eve performs an individual attack, i.e., she attacks each of the states individually, she attaches an ancilla system $\lvert E\rangle_E\langle E\rvert$ to each of the states ρ_A^i. She then applies the same unitary operation U to each composite system, i.e., after this step the ancilla system has the form

$$\rho_E^i = \mathrm{Tr}_A\left(U^\dagger \rho_A^i \otimes \lvert E\rangle_E\langle E\rvert U\right) \tag{5.31}$$

for all states ρ_A^i that Alice sent. Eve then measures her part of all the composite systems individually and with the same POVM $\mathcal{M}$.

Collective attacks are a little more general in the sense that Eve is allowed to measure all states collectively instead of individually, but she is still restricted to attaching individual ancilla systems to each state.

The most general attack is the coherent attack. Here, Eve attaches one large ancilla system to all states that Alice sends, i.e., to the state $\rho_A^1 \otimes \rho_A^2 \otimes \cdots \otimes \rho_A^n$, and applies a global unitary U_G to the whole composite system. This means that the state of the ancilla system before the measurement is

$$\rho_E = \mathrm{Tr}_A\left(U_G^\dagger(\rho_A^1 \otimes \cdots \otimes \rho_A^n) \otimes \lvert E\rangle_E\langle E\rvert U_G\right). \tag{5.32}$$

Similar to collective attacks, Eve then performs a global measurement on her part of the system.

Individual Attacks

Since individual attacks are the ones that can be analysed most easily, we want to have a closer look at the techniques that can be employed here. When doing an individual attack, Eve is restricted to the same interaction with each quantum state sent by Alice, which makes it easy to characterize and analyse the possible attacks. For this purpose we are interested in the amount of information about a single state that Eve gains with an attack. For simplicity, we now only focus on the mutual information between Eve and Alice, following the historical definition of security. Although this alone is not a sufficient security criterion (as discussed above), it already provides a lot of insights on individual attacks and, moreover, it has been studied in detail.

The most general individual attack that Eve can perform on a qubit system (as in the BB84 or in the six-state protocol) is the following: Since every system has to be attacked individually and in the same way, the only degree of freedom lies in the unitary operation that Eve applies to the composite system. The most general unitary transformation is

$$U|0\rangle|E\rangle = \sqrt{F}|0\rangle|E_{00}\rangle + \sqrt{1-F}|1\rangle|E_{01}\rangle \tag{5.33}$$

$$U|1\rangle|E\rangle = \sqrt{F}|1\rangle|E_{10}\rangle + \sqrt{1-F}|0\rangle|E_{11}\rangle, \tag{5.34}$$

where $\{|0\rangle, |1\rangle\}$ is the computational basis for Alice's state, the first qubit is the one Alice sends to Bob, and the second qubit represents Eve's ancilla system, which is in the initial state $|E\rangle$. The states $|E_{00}\rangle$, $|E_{01}\rangle$, $|E_{10}\rangle$, and $|E_{11}\rangle$ represent the state of Eve's ancilla system after the unitary operation. F is the fidelity (as defined in Definition 2.57) between the initial state $|\psi_{\text{in}}\rangle$ sent by Alice and the state ρ_B that Bob receives (which is the right hand side of (5.33) and (5.34) traced over Eve's states), i.e.,

$$F = \langle\psi_{\text{in}}|\rho_B|\psi_{\text{in}}\rangle. \tag{5.35}$$

We assume that Eve is clever enough to treat all employed basis in the protocol in the same way, i.e., the fidelity is the same for all bases. Otherwise, Alice and Bob would be able to detect the eavesdropper by comparing the fidelities for different bases.

For the BB84 protocol, it was shown in [14] that the mutual information between Alice and Bob, $I(A:B)$, and the mutual information between Alice and Eve, $I(A:E)$, can be expressed in terms of the *disturbance* $D = 1 - F$ in the following way:

$$I(A:B) = 1 + D\log D + (1-D)\log(1-D), \tag{5.36}$$

$$I(A:E) = \frac{1}{2}(1 + f(D))\log(1 + f(D)) + \frac{1}{2}(1 - f(D))\log(1 - f(D)), \tag{5.37}$$

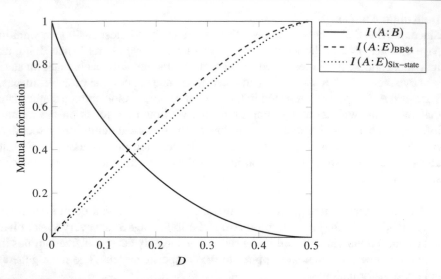

Fig. 5.3 Comparison of mutual information. The plot shows the mutual information between Alice and Bob (solid line), which is the same for the BB84 and the six-state protocol alongside the mutual information between Alice and Eve (dashed for the BB84 protocol, dotted for the six-state protocol)

where $f(D) = 2\sqrt{D(1 - D)}$. The same analysis can be done for the six-state protocol (see [7]), which leads to the same expression for $I(A : B)$, but to a different one for $I(A : E)$:

$$I(A : E) = 1 + (1 - D)\Big(g(D) \log g(D) + (1 - g(D)) \log(1 - g(D))\Big),$$

$$(5.38)$$

where $g(D) = \frac{1}{2}\Big(1 + \frac{1}{1-D}\sqrt{D(2 - 3D)}\Big)$. In Fig. 5.3, the mutual information between Alice and Bob $I(A : B)$ is depicted alongside the mutual information between Alice and Eve, $I(A : E)$, for the BB84 and six-state protocols. According to the Csiszár-Körner theorem [12], Alice and Bob can distil a secret key whenever they have an advantage over Eve in terms of the mutual information. To illustrate this, the difference between the Alice's and Bob's mutual information and the Alice's and Eve's mutual information, $I(A : B) - I(A : E)$, is plotted in Fig. 5.4 for the BB84 and six-state protocols. When this quantity becomes negative, no secret key can be distilled.

Collective and Coherent Attacks

More sophisticated attacks, like collective and coherent attacks, are much harder to analyse than individual ones. Especially in case of coherent attacks, the corresponding global Hilbert space dimension is very high since Eve interacts with all the states simultaneously. In terms of security proofs, several techniques have

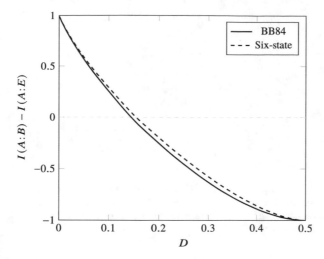

Fig. 5.4 Difference of mutual informations. In this plot, the difference between Alice's and Bob's mutual information and Alice's and Eve's mutual information is depicted, both for the BB84 protocol (solid) and the six-state protocol (dashed)

been developed to prove security against general attacks. The key ingredients here are often theorems that reduce the security proof against coherent attacks to a security proof against collective attacks, such as quantum de Finetti theorems or the postselection technique. We go into more detail about techniques for security proofs later in this chapter.

For the BB84 protocol, coherent eavesdropping attacks have been studied in [10], and for the six-state protocol they have been studied in [2]. In brief, the authors find that coherent eavesdropping does not help Eve to obtain more information, but it increases the probability to correctly guess the whole message sent by Alice.

5.3.2 Photon-Number-Splitting Attack

The photon-number-splitting attack (depicted in Fig. 5.5) is an attack against a certain implementation of a single-photon source using weak laser pulses, which we have already briefly mentioned in Sect. 4.1.1 when we discussed the SARG04 protocol. It exploits the fact that in practical QKD schemes, qubits are usually represented by photons, using the polarization as a degree of freedom. Ideally, one qubit is encoded by exactly one photon. However, in practice, ideal single-photon sources do not exist. Therefore, one often uses *weak coherent laser pulses* instead. A coherent state with (complex) phase α is given by

$$|\alpha\rangle = e^{\frac{|\alpha|^2}{2}} \sum_{n=0}^{\infty} \frac{\alpha^n}{\sqrt{n!}} |n\rangle, \tag{5.39}$$

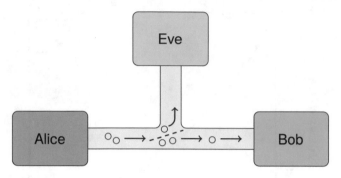

Fig. 5.5 The photon-number-splitting (PNS) attack. Whenever the signal sent by Alice contains more than one photon, Eve splits off one of them and stores it in her quantum memory. After Alice has announced her encoding bases, Eve measures these photons in the correct basis and gets perfect information about these key bits

where $|n\rangle$ is a Fock state that represents the quantum state when the pulse contains n photons. If the phase α is unknown or randomized, we get *phase-randomized coherent states*:

$$\rho = \int \frac{d \arg \alpha}{2\pi} |\alpha\rangle\langle\alpha| = \sum_{n=0}^{\infty} P(n)|n\rangle\langle n|, \qquad (5.40)$$

where $P(n)$ is the probability distribution of the number of photons, which is given by the Poisson distribution

$$P(n) = e^{-|\alpha|^2} \frac{|\alpha|^{2n}}{n!}. \qquad (5.41)$$

Here, $|\alpha|^2 = \mu$ describes the average photon number of the signals. Hence, the state that Alice sends to Bob is given by

$$\rho = e^{-\mu} \sum_{n=0}^{\infty} \frac{\mu^n}{n!} |n\rangle\langle n|, \qquad (5.42)$$

and the information is imprinted in the polarization of these photons. A typical weak laser pulse has an average photon number of $\mu = 0.1$. In this case, most of the signals are vacuum signals: The probability that no photon is sent is given by $P(0) = e^{-\mu} \approx 90.5\%$. The event that exactly one photon is sent happens with probability $P(1) = \mu e^{-\mu} \approx 9\%$, and the event that more multiple photons are sent happens with probability $P(n > 1) = 1 - (1 + \mu)e^{-\mu} \approx 0.5\%$.

Let us now consider what happens if we use this realistic photon source instead of an ideal single-photon source in the BB84 protocol. The cases where no photon is sent simply reduce the signal rate, since Bob does not detect anything, but no information is revealed to Eve. The single-photon signals work exactly as signals from an ideal single-photon source. The only problem are the multi-photon pulses. Eve can exploit the existence of these signals to perform the *photon-number-splitting* attack, which is especially useful if Alice and Bob use a lossy quantum channel (which is usually the case in practical implementations).

It works as follows: suppose Alice and Bob have chosen to use weak pulses with an average photon number μ. The quality of the channel is characterized by its single-photon transmittance (which is the probability that a single photon is transmitted by the channel and does not get lost), denoted as η. Hence, Bob expects to receive states distributed by a Poisson distribution with average detected photon number $\mu\eta$ and therefore expects to receive non-vacuum events with probability

$$P_{\text{non-vac}} = 1 - e^{-\mu\eta}. \tag{5.43}$$

For Eve to remain undetected, it is important that Bob receives the correct fraction of no detection and detection events. Eve's strategy is then the following: she replaces the lossy channel by a perfect quantum channel. Then she performs a quantum non-demolition measurement on each pulse, which tells her the exact number of photons of the pulse. Such a measurement can be done without disturbing the polarization of the photons. Eve then guides the signals depending on their number of photons (also depicted in Fig. 5.6):

1. Vacuum events are simply forwarded to Bob, since Eve cannot learn anything about the polarization from them.
2. From multi-photon signals she splits off one photon and forwards the remaining photons to Bob without disturbing the polarization either in the photon she splits off or in the photons she forwards to Bob (depicted in Fig. 5.5). Later in the protocol, when Alice reveals the polarization bases she choses for each signal,

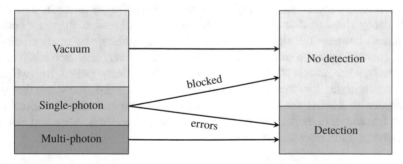

Fig. 5.6 Eve's strategy in the PNS attack. Depending on the number of photons, Eve guides the signals sent by Alice differently: Vacuum signals are passed to Bob, leading to no detection events. From all multi-photon signals, Eve splits off a single photon and passes the remaining signal to Bob, leading to detection events. Single-photon signals can either be blocked to achieve the correct fraction of detection and no detection events that Bob expects. With remaining single-photons events, Eve can perform any coherent eavesdropping attack that possibly induces errors

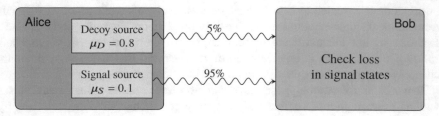

Fig. 5.7 Example of the decoy state strategy. Alice uses a second source of weak laser pulses with a higher average photon number (e.g., $\mu_D = 0.8$ and $\mu_S = 0.1$), the decoy state source. She then randomly sends decoy states to Bob in between the signal states, for instance with a probability of 5%. After Bob has received the states, Alice announces which states have been decoy states and Bob checks the loss in the signal states. If Eve has performed a PNS attack, he will observe a much higher loss in those states than expected

Eve can perform the correct measurement on these photons and gain perfect information about the encoded bits.

3. A fraction of the single-photon events is blocked in order to match Bob's expectation of receiving detection events with probability $1 - e^{-\mu\eta}$. On the remaining single-photon signals, which are not blocked, Eve can perform any coherent attack she likes, thereby possibly introducing errors to the state.

How can Alice and Bob protect themselves against this attack? We have already seen one possibility, the SARG04 protocol. Here, they use a different sifting technique so that Alice never has to reveal her choice of basis. Another strategy, which was proposed in [16, 19, 33], is using the so-called *decoy states*. The idea is to employ a second photon source in Alice's lab that emits weak coherent laser pulses with a different photon number distribution (the decoy states), see Fig. 5.7. More precisely, the additional source has a much higher average photon number μ than the source used to send the actual signals, but otherwise it does not differ from the signal source in terms of other parameters like wavelength, etc. Alice sends decoy states to Bob at random between the signal pulses. Since Eve cannot distinguish between the signal pulses and the decoy pulses, she treats all signals equally according to the strategy described above. However, after Bob has received the signals, Alice reveals which of the pulses have been decoy pulses. Bob will then find a much higher loss than expected in the signal states (i.e., the states that were prepared with a lower mean photon number), since Eve has tried to maintain the wrong fraction of detection and no detection events. Therefore, Alice and Bob can detect the eavesdropper with this strategy.

5.4   Security of BB84

In the paper that originally introduced the BB84 protocol [4] the authors proved it to be secure against certain attacks. However, it took a remarkable amount of time until the security of the BB84 protocol against an adversary that was only limited by the laws of physics could be rigorously proven. At the end of the 1990s, several proofs have been presented: in [18], the security was proven in a way that required perfect quantum computation, which makes it infeasible with current technology.[8] Two other proofs have been given in [22] and [6], which do not require quantum computation but are somewhat complicated. In [30], Shor and Preskill presented a simple proof of the security of the BB84 protocol that builds on ideas of the previous proofs. In this section, we will follow ideas from this work and the presentation of this proof in [23] and [8].

Even though the proof is very specific to the states and measurements of the BB84 protocol and therefore difficult to apply to other protocols we present the proof here in detail since it was the first complete security proof of a QKD protocol and is therefore of interest for historical reasons, but it also illustrates the way people thought about security at the beginning of the 2000s. Later in this chapter we discuss modern security techniques that do not rely as heavily on the specifics of the protocol and are therefore applicable to a variety of protocols.

The structure of the security proof of the BB84 protocol is as follows: first, we prove the security of the entanglement-based version of the BB84 protocol. Then we show how this protocol can be reduced to the original prepare-and-measure version of BB84.

5.4.1   Security of the Entanglement-Based Version

We begin the security analysis of the BB84 protocol with the entanglement-based version of it that was introduced in the previous chapter. Recall that the main goal of this protocol is to establish shared maximally entangled states between Alice and Bob, namely the Bell state $|\Phi^+\rangle^{\otimes m}$, where

$$|\Phi^+\rangle = \frac{1}{\sqrt{2}}\big(|00\rangle + |11\rangle\big). \tag{5.44}$$

However, due to losses in the quantum channel and Eve's interaction with the states, Alice and Bob will not end up with exactly this state, but with a state ρ_{AB} that is (hopefully) similar to $|\Phi^+\rangle^{\otimes m}$. As in the previous section, we measure the closeness of these two states in terms of the fidelity:

$$F\big(\rho_{AB}, |\Phi^+\rangle\big) = \langle\Phi^+|\rho_{AB}|\Phi^+\rangle. \tag{5.45}$$

If the fidelity is 1, the state that Alice and Bob share is equal to the maximally entangled state $|\Phi^+\rangle$. In this case, they can be sure that Eve's system is completely uncorrelated to theirs, since a maximally entangled state between two parties cannot

---

[8]This work is based on a slightly different version of the entanglement-based BB84 protocol where, instead of using CSS codes, quantum computers are used for error correction.

have any correlation with a third party. Alice and Bob can then safely measure this shared state in order to generate a secret key (i.e., the protocol is secure). However, if their shared state is not equal but only close to a maximally entangled state, it is possible that Eve's system is correlated with that state, which means that she can gain some information on the key. This naturally leads to the question what impact a fidelity below 1 has on the security of the protocol. We use the following lemma (see [17]) to connect the fidelity to the security of the protocol:

Lemma 5.6 *Let $\epsilon \geq 0$ and ρ_{AB} be a bipartite quantum state such that*

$$F\left(\rho_{AB}, |\Phi^+\rangle^{\otimes m}\right) \geq 1 - \epsilon^2. \tag{5.46}$$

Then the two n-bit strings resulting from local measurements of ρ_{AB} in the computational basis are ϵ-secure keys (with respect to an adversary who controls a purifying system of ρ_{AB}.)

A short note before we give the proof of this lemma: the term *purifying system* has the following meaning: since we do not impose any restrictions on Eve's attacks, we always assume the worst case. The worst case here means that the composite system of Alice, Bob, and Eve is in a pure state $|\psi_{ABE}\rangle$ such that $\rho_{AB} = \text{Tr}_E\left(|\psi_{ABE}\rangle\langle\psi_{ABE}|\right)$ and $\rho_E = \text{Tr}_{AB}\left(|\psi_{ABE}\rangle\langle\psi_{ABE}|\right)$. This state is called the *purification*[9] of ρ_{AB} and it corresponds to the scenario where Eve has the most power because every other extension ρ_{ABE} of ρ_{AB} can be obtained from the pure state $|\psi_{ABE}\rangle$ by doing a quantum operation on Eve's system.

Proof Recall that, in order to prove that a protocol is secure, we must show that

$$\frac{1}{2}\|\rho_{ABE} - \rho_{UU} \otimes \rho_E\|_1 \leq \epsilon, \tag{5.47}$$

where $\rho_{UU} = \sum_{u \in S} \frac{1}{|S|} |u\rangle\langle u| \otimes |u\rangle\langle u|$, which represents a key that is completely random but the same for both systems. By Uhlmann's theorem (see Theorem 2.61), there is a quantum state σ_E of Eve's system such that

$$F\left(\rho_{ABE}, |\Phi^+\rangle^{\otimes m} \otimes \sigma_E\right) = F\left(\rho_{AB}, |\Phi^+\rangle^{\otimes m}\right) \tag{5.48}$$

---

[9]Every density operator ρ has a purification on some higher-dimensional Hilbert space. Since we do not impose any restrictions on Eve's Hilbert space, it can be big enough to admit a purification of any state ρ_{AB} that Alice and Bob share. A detailed treatment of this concept can be found in [34].

(note that the state ρ_{ABE} is a purification of ρ_{AB}). We now use the relation between the trace distance and the fidelity given in Theorem 2.59:

$$\frac{1}{2}||\rho_{ABE} - |\Phi^+\rangle^{\otimes m} \otimes \sigma_E||_1 \leq \sqrt{1 - F(\rho_{ABE}, |\Phi^+\rangle^{\otimes m} \otimes \sigma_E)}$$

$$= \sqrt{1 - F(\rho_{AB}, |\Phi^+\rangle^{\otimes m})}$$

$$\leq \sqrt{1 - (1 - \epsilon^2)}$$

$$= \epsilon.$$

We have not quite arrived at the security condition (5.47) yet: The last ingredient we need is the observation that Alice and Bob can transform the shared maximally entangled state $|\Phi^+\rangle^{\otimes m}$ into a state of the form ρ_{UU} by simply measuring it in the computational basis. Since their respective qubits are perfectly correlated, their measurement results are always the same, and because of the structure of the state each result is obtained with probability $\frac{1}{2}$. □

With the above lemma at hand, it now remains to show how Alice and Bob can estimate the fidelity. In practice, Alice and Bob cannot estimate the fidelity of the key bits directly since this includes measuring them, which destroys the state. However, Alice and Bob can make statements about the statistics of the remaining states if they choose the set of check states large enough, as we have seen in Sect. 4.2.1 when discussing parameter estimation. Since the theorem we used in the analysis of parameter estimation, Theorem 4.5, is an argument based on classical probability, it is not immediately clear that we can apply it to the outcomes of quantum measurements.[10]

Fortunately, quantum measurements do allow an interpretation in terms of classical probability theory if the observables that are considered refer to only one basis. This is indeed the case for the present situation, since we can make use of the *Bell basis*, which consists of the four states

$$|\Phi^+\rangle = \frac{|00\rangle + |11\rangle}{\sqrt{2}}, \tag{5.49}$$

$$|\Phi^-\rangle = \frac{|00\rangle - |11\rangle}{\sqrt{2}}, \tag{5.50}$$

$$|\Psi^+\rangle = \frac{|01\rangle + |10\rangle}{\sqrt{2}}, \tag{5.51}$$

$$|\Psi^-\rangle = \frac{|01\rangle - |10\rangle}{\sqrt{2}}. \tag{5.52}$$

---

[10] A very good demonstration of the problems that can arise here is the CHSH inequality described in the previous chapter, where the outcomes of a quantum measurement violate the bound that is derived using classical probability theory.

In the protocol, Alice holds the state $|\Phi^+\rangle^{\otimes m}$ and sends the second qubit of all the individual states to Bob. In general, there are three different errors a qubit can undergo: a bit flip, a phase error, and a combination of the two. A bit flip changes $|0\rangle$ to $|1\rangle$ and $|1\rangle$ to $|0\rangle$. A phase error leaves the state $|0\rangle$ unchanged but transforms $|1\rangle$ into $-|1\rangle$. The combination of both errors corresponds to the transformation $|0\rangle \rightarrow -|1\rangle$ and $|1\rangle \rightarrow |0\rangle$. These three errors are generated by the Pauli matrices

$$\sigma_x = \begin{pmatrix} 0 & 1 \\ 1 & 0 \end{pmatrix}, \quad \sigma_y = \begin{pmatrix} 0 & -i \\ i & 0 \end{pmatrix}, \quad \sigma_z = \begin{pmatrix} 1 & 0 \\ 0 & -1 \end{pmatrix}, \tag{5.53}$$

where the bit flip error corresponds to σ_x, the phase error corresponds to σ_z, and the combination of both to σ_y. Therefore, the initial state that Alice holds $|\Phi^+\rangle$ can undergo four different evolutions when she sends the second qubit to Bob (note that we ignore all irrelevant overall phases):

1. Nothing happens to the qubit (which corresponds to applying the identity operator) and the state after the transmission is still $|\Phi^+\rangle$.
2. A bit flip error occurs (which corresponds to applying the σ_x operator to the qubit). This will transform the state into $|\Psi^+\rangle$.
3. A phase error occurs (which corresponds to applying the σ_z operator), which transforms the state into $|\Phi^-\rangle$.
4. A combination of bit flip and phase error occurs (which corresponds to applying the σ_y operator), which transforms the state into $|\Psi^-\rangle$.

These errors can be detected by the following measurements: the POVM that describes the measurement to detect a bit flip error is given by $\{\Pi_{bf}, 1 - \Pi_{bf}\}$ with

$$\Pi_{bf} = |\Psi^+\rangle\langle\Psi^+| + |\Psi^-\rangle\langle\Psi^-|. \tag{5.54}$$

Similarly, to detect a phase error, the measurement is described by the POVM $\{\Pi_{pe}, 1 - \Pi_{pe}\}$, where

$$\Pi_{pe} = |\Phi^-\rangle\langle\Phi^-| + |\Psi^-\rangle\langle\Psi^-|. \tag{5.55}$$

These measurements are given solely in terms of the Bell basis; hence, their outcomes obey classical probability arguments. Although this looks promising, there is still one caveat: The Bell states are non-local, and therefore, measurements in this basis require, in general, non-local operations. These can be difficult to implement since Alice and Bob do not have access to the other party's respective lab. Fortunately, in the present scheme the measurements can be implemented locally since

$$\Pi_{bf} = \frac{1}{2}\left(\text{id} \otimes \text{id} - \sigma_z \otimes \sigma_z\right), \tag{5.56}$$

$$\Pi_{pe} = \frac{1}{2}\left(\text{id} \otimes \text{id} - \sigma_x \otimes \sigma_x\right). \tag{5.57}$$

Therefore, Alice and Bob can perform the desired checks with only local operations and estimate the fidelity in this way.

Exercise 5.7 Show that the measurement operators given in (5.56) and (5.57) yield the same outcome statistics as the ones defined in (5.54) and (5.55).

5.4.2 Reduction to the Prepare-and-Measure Version

To achieve our ultimate goal, namely showing that the original version of the BB84 protocol is secure, we still need to show that the entanglement-based version of the BB84 protocol is equivalent to its prepare-and-measure version. For this purpose, we begin with the scheme given in the entanglement-based BB84 protocol and systematically modify different steps to finally arrive at the original version. In this part of the proof we use several aspects of the construction of classical linear codes and CSS codes as well as properties of cosets; hence, it is helpful to have a look at Appendix B first in order to follow the arguments below.

We begin the reduction by removing the need to distribute the maximally entangled pairs $|\Phi^+\rangle^{\otimes 2n}$. First, note that the measurement of the n check pairs in step 8 is simply used to estimate the error rate. Hence, instead of using entangled states, Alice can equivalently prepare and send n single qubit states to Bob. This changes steps 1, 2, and 8 of the protocol to:

1'. Alice creates n random check bits and n maximally entangled states $|\Phi^+\rangle^{\otimes n}$. She encodes n qubits as $|0\rangle$ or $|1\rangle$ according to the check bits.
2'. She randomly chooses n out of $2n$ positions to put in the check qubits. In the remaining positions she puts in one half of each state $|\Phi^+\rangle$.
8'. Bob measures the n check qubits in the $\{|0\rangle, |1\rangle\}$ basis and publicly shares the result with Alice. If more than t bits disagree, they abort the protocol.

To remove the remaining n entangled pairs, recall some basics of the CSS codes we are using:[11] Given two classical linear error correction codes C_1 (that encodes k_1 bits into n bits) and C_2 (that encodes k_2 bits into n bits), a $[n, m]$ CSS code of C_1 over C_2, denoted as $\mathrm{CSS}(C_1, C_2)$, encodes $m = k_1 - k_2$ qubits into n qubits and corrects up to t errors. To find out the positions of bit flip errors, one has to apply the parity check matrix H_1 of the classical code C_1. To get information about the phase errors, one applies the parity check matrix $H_2^\perp$ of the classical dual code $C_2^\perp$. A codeword in this code is always of the form

$$|x_k + C_2\rangle = \frac{1}{\sqrt{|C_2|}} \sum_{y \in C_2} |x_k + y\rangle, \tag{5.58}$$

---

[11]For more details about CSS codes, see Appendix B.

where x_k is a representative of one of the 2^m cosets of C_2 in C_1. The notation x_k implies that it is a vector x that is indexed by a string k. Furthermore, there exists a family of $[n, m]$ CSS codes of C_1 over C_2 that is equivalent to this one in the sense that these codes can correct the same number of errors, which we denote $\text{CSS}_{v,w}(C_1, C_2)$, where v and w are n-bit strings. The codewords in this family are given by

$$|x_k, v, w\rangle = \frac{1}{\sqrt{|C_2|}} \sum_{y \in C_2} (-1)^{v \cdot y} |x_k + y + w\rangle. \tag{5.59}$$

These states have the important property that they form an orthonormal basis of a 2^n-dimensional Hilbert space. To see this, we will show that there are $2^{k_1 - k_2}$ distinct values of x_k, 2^{n-k_1} distinct values of w, and 2^{k_2} distinct values of v:

i. As discussed in Appendix B, if $x_k - x_k' \in C_2$, then $|x_k + C_2\rangle = |x_k' + C_2\rangle$, which implies that the state $|x_k + C_2\rangle$ only depends on the coset of C_1/C_2 in which x_k is contained. Since there are 2^m such cosets, there are 2^m distinct values of x_k.

ii. Suppose that $|x_k, v, w\rangle = |x_k, v', w\rangle$. This implies that $v \cdot y = v' \cdot Y$ for all $y \in C_2$, and therefore $(v - v') \cdot y = 0$ for all $y \in C_2$. This in turn means that $v - v' \in K$, where K is the row space of the parity check matrix H_2 of C_2 (the rows of H_2 span the space of all vectors that are orthogonal to the codewords y of C_2). Therefore, the requirement for two states $|x_k, v, w\rangle$ and $|x_k, v', w\rangle$ to be distinct states is $v - v' \notin K$, which directly implies $v + K \neq v' + K$ (which is a property of cosets). Since H_2 has $n - k_2$ linearly independent rows, v has $2^{n-(n-k_2)} = 2^{k_2}$ distinct values.

iii. With a similar argument as in 1., the values of w depend on the coset of C_1 in a 2^n-dimensional Hilbert space in which w is contained, which implies that there are 2^{n-k_1} distinct values of w.

It remains to show that the states are orthonormal: Note that if x_k and x_k' belong to different cosets of C_1/C_2, then there are no codewords $y, y' \in C_2$ such that $x_k + y = x_k' + y'$; hence, the states in (5.58) are orthonormal states. A similar argument can be applied to the current situation: Consider the cosets of C_2 in $\mathbb{F}_2^n$ (the set of all n-bit strings). For two distinct states $|x_k, v, w\rangle$ and $|x_k', v', w'\rangle$, $x_k + w$ and $x_k' + w$ belong to different cosets of $\mathbb{F}_2^n/C_2$ and therefore there are no $y, y' \in C_2$ such that $x_k + w + y = x_k' + w' + y'$; hence, the two states are orthonormal.

We can use the fact that the states defined in (5.59) form an orthonormal basis of a 2^n-dimensional Hilbert space to express the maximally entangled pairs that Alice and Bob share in this basis:

$$|\Phi^+\rangle^{\otimes n} = \frac{1}{\sqrt{2^n}} \sum_{j=0}^{2^n-1} |j\rangle|j\rangle = \frac{1}{\sqrt{2^n}} \sum_{x_k, v, w} |x_k, v, w\rangle|x_k, v, w\rangle, \tag{5.60}$$

where the first state corresponds to Alice's qubits and the second one is the qubit she sends to Bob. When Alice measures the error syndrome corresponding to H_1 and $H_2^\perp$ on her qubits in step 9, she obtains random values for v and w. Similarly, in step 10, the measurement in the computational basis yields a random string $x_k \in C_1/C_2$. Therefore, Bob's state has collapsed to $|x_k, v, w\rangle$, which is the codeword for x_k in $\mathrm{CSS}_{v,w}(C_1, C_2)$. This is the encoded counterpart of an m-qubit state $|k\rangle$ (remember that the codes in this family encode m qubits into n qubits). It follows that instead of preparing maximally entangled states, Alice can equivalently choose bit strings v, w and k at random, encode $|k\rangle$ in the code $\mathrm{CSS}_{v,w}(C_1, C_2)$, and send the corresponding n-qubit codeword to Bob. The resulting scheme, called the *CSS codes* protocol, is then as follows:

CSS codes protocol

1. Alice creates n random check bits, a random m-bit key k, and two random n-bit strings v and w. She encodes $|k\rangle$ in the code $\mathrm{CSS}_{v,w}(C_1, C_2)$. She also encodes n qubits as $|0\rangle$ or $|1\rangle$ according to the check bits.
2. She randomly selects n positions out of $2n$ and puts the check qubits at these positions and the encoded qubits in the remaining positions.
3. Alice selects a random classical bit string $b = (b_1, b_2, \ldots, b_{2n})$ of length $2n$. Whenever the bit b_i is 1, she applies a Hadamard transformation (2.28) to her half of the corresponding qubit pair.
4. She sends the other half of all qubit pairs to Bob.
5. Bob receives the qubits and publicly announces this fact.
6. Alice announces b, v and w and the positions of the check qubits.
7. Bob applies a Hadamard transformation to those qubits for which $b_i = 1$.
8. Bob measures the n check qubits in the computational basis $\{|0\rangle, |1\rangle\}$ to estimate the error rate. If more than t errors occur, they abort the protocol.
9. If the number of errors is below t, Bob decodes the remaining n qubits from $\mathrm{CSS}_{v,w}(C_1, C_2)$.
10. Bob measures his qubits to obtain the shared secret key k.

The CSS codes protocol is secure because it was derived by reduction from the entanglement-based version of BB84, which we proved to be secure in the previous section. However, we have not yet arrived at the prepare-and-measure version of BB84. The CSS codes protocol still requires perfect quantum computation for the encoding and decoding and furthermore, Bob needs quantum memory for the storage of his qubits, while he waits for Alice's announcements. In contrast, the prepare-and-measure version of the BB84 protocol only requires single qubit preparation and measurements. However, the use of CSS codes makes it possible to remove these requirements because of their property that they decouple bit flip and phase errors.

First, note that Bob only cares about the bit values of the encoded key and the string v is only needed to correct the phase of the encoded qubits; hence, Bob does not need v. Since Alice does not need to reveal v, she effectively sends a mixed state averaged over all values of v:

$$\rho_{x_k, w} = \frac{1}{2^n} \sum_v |x_k, v, w\rangle \langle x_k, v, w| \tag{5.61}$$

$$= \frac{1}{2^n |C_2|} \sum_v \sum_{y, y' \in C_2} (-1)^{v \cdot (y - y')} |x_k + y + w\rangle \langle x_k + y' + w| \tag{5.62}$$

$$= \frac{1}{|C_2|} \sum_{y \in C_2} |x_k + y + w\rangle \langle x_k + y + w|. \tag{5.63}$$

Equivalently, this state can be prepared using only single qubit operations by choosing a codeword. Therefore, the first step of the CSS codes protocol is modified to:

1'. Alice creates n random check bits, a random n-bit string w, a random codeword $x_k \in C_1/C_2$, and a random codeword $y \in C_2$. She encodes n qubits in the state $|0\rangle$ or $|1\rangle$ according to $x_k + y + w$, and n qubits in $|0\rangle$ or $|1\rangle$ according to the check bits.

Bob will receive a state $|x_k + y + w + e\rangle$ and he can, instead of first decoding the state and then measuring it as described in steps 9 and 10, directly measure it in the computational basis. Then he obtains a string $x_k + y + w + e$ and now he can do the error correction *classically* since C_1 and C_2 are classical codes. With the error correction w he has received from Alice, he can subtract w from his string and obtain $x_k + y + e$. If e does not contain too many errors, Bob can then correct $x_k + y + e$ unambiguously to the codeword $x_k + y$.

This scheme becomes even more simple if Alice chooses $x_k \in C_1$ instead of $x_k \in C_1/C_2$, because y is not necessary any more. Moreover, $x_k + w$ is a completely random n-bit string, which means that Alice can equivalently choose a string x at random and send the state $|x\rangle$ to Bob, who measures it to obtain $x + e$. Alice then sends the string $x - x_k$, so Bob can subtract this from his string to get $x_k + e$, which he corrects to x_k. We have achieved now that the key qubits are prepared in a random state, in the same way as the check bits! This results in the following modified steps:

1''. Alice chooses a random $x_k \in C_1$ and creates a random $2n$-bit string, according to which she creates $2n$ qubits in the $|0\rangle$ or $|1\rangle$ state.
2'. Alice randomly chooses n out of $2n$ positions to be the check qubits and the remaining n positions to define $|x\rangle$.
6'. Alice announces b, $x - x_k$, and the positions of the n check bits.

9'. Bob measures the remaining qubits to obtain $x + e$, subtracts $x - x_k$ from this, and corrects it with the code C_1 to obtain x_k.

10'. Alice and Bob compute the coset to which x_k belongs in order to obtain the final key k.

We can also remove the Hadamard transformation from the protocol. Alice can instead encode her qubits directly either in the computational basis $\{|0\rangle, |1\rangle\}$ or in the Hadamard basis $\{|+\rangle, |-\rangle\}$, depending on the values of the bits in b. Bob then directly measures these qubits at random either in the computational or in the Hadamard basis. When Alice announces b, they will only keep those bits for which they have chosen the same basis. This removes the requirement for Bob to have a quantum memory to store the qubits until he receives information from Alice. However, since they discard with high probability about half of their bits in this step, they have to start with $4n$ qubits in order to get the same amount of key in the end.

Because of the sifting step we have added, Alice has to wait with her choice of check qubits until after the bits have been discarded. With these last modifications, we arrive at the final protocol, which is the same as the prepare-and-measure version of BB84:

Secure BB84

1. Alice creates $4n$ random bits.
2. Alice encodes each of the bits either in the computational or in the Hadamard basis according to another random $4n$-bit string b.
3. Alice sends the resulting qubits to Bob.
4. Alice chooses a random $x_k \in C_1$.
5. Bob receives the qubits, publicly announces this fact, and measures each of the qubits either in the computational or in the Hadamard basis, chosen at random.
6. Alice announces b.
7. Alice and Bob discard those bits where they have used different bases. After this step, there are about $2n$ bits left. Alice randomly chooses n of these bits to serve as check bits and tells Bob the position of these bits.
8. Alice and Bob publicly compare the check bits. If they find more than t errors, they abort the protocol. Otherwise, they continue and Alice is left with the n-bit string x, and Bob with n-bit string $x + e$.
9. Alice announces $x - x_k$. Bob subtracts this from his string and corrects it with the code C_1 to obtain x_k.
10. Alice and Bob compute the coset to which x_k belongs in order to obtain the final key k.

Note that step 9 serves as the error correction step since C_1 is simply a classical error correction code. Step 10 is privacy amplification: we assume that Eve has no knowledge about the code C_1 (which can be achieved by Alice choosing the code at random); hence, even if she has some knowledge about x_k, without knowing the code C_1, she does not know which coset x_k belongs to.

We have now proven the security of the original BB84 protocol by first proving the security of its entanglement-based version and then successively reducing that scheme to the prepare-and-measure version. Since we did not make any modifications on Eve's quantum state, we conclude that the BB84 protocol is secure.

However, there are of course some caveats: This proof only proves the security of the ideal protocol, where the states sent are exactly those described. This does not guarantee security against the photon-number-splitting attack, for example. Furthermore, the proof does not make any statements about the amount of effort needed for the decoding: for practical applications in QKD, the code C_1 must be efficiently decodable. Another point is that the proof does not provide an upper bound for the amount of eavesdropping that is tolerable, since CSS codes are not optimal. In [30], it is estimated that a rate of bit flip and phase errors of up to 11% is tolerable using a protocol similar to the BB84 protocol. However, with the aid of quantum computation for the encoding and decoding, it might be possible to tolerate higher error rates.

5.5   Modern Techniques

The security proof of BB84 presented above was a remarkable achievement since it took many years from the first presentation of the protocol until its security could be proven. However, the proof does not have a modular structure in the sense that it mixes different steps of the protocol such as error correction and privacy amplification. This makes it difficult to transfer the techniques used there to other protocols since they are very specific to the states and measurements used in the BB84 protocol. Therefore, other techniques have been developed since the days of the first security proofs, which are more general and can be adapted to a variety of protocols. In the following, we discuss the most important state-of-the-art techniques in security proofs.

5.5.1   The Secret Key Rate

One of the most important quantities for modern security proofs is the *Devetak–Winter rate*, which gives a lower bound on the asymptotic *secret key rate* r.[12] The

---

[12]Asymptotic means that the bound holds for an infinite number of repetitions of the experiment. While this is an unrealistic assumption in practice (since it does not account for any finite-size effects), it is usually easier to analyse. We discuss the finite case later in this chapter.

secret key rate is the number of secure key bits that can be extracted per signal sent. In [13], Devetak and Winter prove a lower bound on this quantity for the case of collective attacks and in the regime of an infinite key: let K_A and K_B be the random variables that describe Alice's and Bob's keys, respectively, and E the quantum system that Eve holds. The asymptotic secret key rate is then lower bounded by

$$r \geq I(A : B) - I(A : E). \tag{5.64}$$

An intuitive interpretation of this definition is as follows: the fraction of secret bits generated per round of using the protocol is equal to the amount of information shared by Alice and Bob, $I(A : B)$, minus the amount of information that Eve has on Alice's part of the key, $I(A : E)$.

Even though the bound on the secret key rate in (5.69) only holds for the case of collective attacks, it is possible to infer security against general attacks from security against collective attacks. One tool that allows for this is the *postselection technique* [9]. This can be applied to protocols where the dimension $d = \dim(\mathcal{H}_A \otimes \mathcal{H}_B)$ of Alice's and Bob's quantum system is known. The statement of the technique is the following: If a QKD protocol of M rounds is ϵ-secure against collective attacks, then it is also $(M + 1)^{d^2-1}\epsilon$-secure against coherent attacks if the length of the secure key is shortened by $2(d^2 - 1) \log(M + 1)$ bits [15].

5.5.2 Security from Entropic Uncertainty Relations

One possibility to evaluate the right hand side of (5.69) is to exploit the explicit form of the protocol in order to reduce the entropy that involves Eve's system to quantities that only depend on Alice's and Bob's system. Another way to find a bound on the secret key rate is using *entropic uncertainty relations*, which we have already seen in Sect. 3.3. Since they constrain the potential knowledge that one can have on a physical system, we can use them to bound Eve's knowledge on the key, following the presentation in [5].

Consider the following situation as it usually occurs in an entanglement-based protocol: The eavesdropper creates a quantum state ρ_{ABE} and distributes part A of the state to Alice and part B to Bob. Of course, in practice Alice and Bob do not provide Eve with such power; however, a security proof that covers this extreme situation obviously includes the case where Alice and Bob distribute the states themselves. To generate the key, Alice and Bob measure the states they receive choosing between two different measurements at random. We denote Alice's possible measurements R and S, and Bob's possible measurements R' and S'. After the measurement, Alice and Bob compare their choices in order to ensure that the generated keys are identical. This communication can entirely be overheard by the eavesdropper who is trying to get information on the key. However, if the measurement outcomes are sufficiently correlated, Alice and Bob can still generate a secure key.

To see this, first recall from Theorem 3.54 the relation

$$H(R|B) + H(S|B) \geq \log\left(\frac{1}{c}\right) + H(A|B). \tag{5.65}$$

Using this relation, we can prove the following corollary, which was first conjectured by [25]:

Corollary 5.8 *For any density operator $\rho_{ABE} \in \mathcal{B}(\mathcal{H}_A \otimes \mathcal{H}_B \otimes \mathcal{H}_E)$,*

$$H(R|E) + H(S|B) \geq \log\left(\frac{1}{c}\right). \tag{5.66}$$

Proof Using the definition of the conditional entropy as given in Definition 3.31, we know that $H(R|B) = H(RB) - H(B)$. Using this, we can rewrite (5.65) as

$$H(RB) + H(SB) \geq \log\left(\frac{1}{c}\right) + H(AB) + H(B). \tag{5.67}$$

If ρ_{ABE} is a pure state, we know from Theorem 3.28 that for any bipartition, the marginal entropies are equal, i.e., $H(RB) = H(RE)$ and $H(AB) = H(E)$. Therefore, we have

$$H(RE) + H(SB) \geq \log\left(\frac{1}{c}\right) + H(E) + H(B), \tag{5.68}$$

which is equivalent to the statement of the corollary. For general states ρ_{ABE}, the result follows by the concavity of the conditional entropy.   □

With this result we can now bound the secret key rate r, using an alternative formulation for the lower bound:

$$r \geq H(R|E) - H(R|B). \tag{5.69}$$

Using (5.69) and Corollary 5.8, we can lower bound the key rate by

$$r \geq \log\left(\frac{1}{c}\right) - H(R|B) - H(S|B) \tag{5.70}$$

$$\geq \log\left(\frac{1}{c}\right) - H(R|R') - H(S|S'), \tag{5.71}$$

where we have used that measurements cannot decrease the conditional entropy (see Theorem 3.35) in the last step. This security argument has the advantage that Alice and Bob only need to bound the entropies $H(R|R')$ and $H(S|S')$, which are both directly observable quantities.

The argumentation has the limitation that it only applies to collective attacks, because the bound derived in [13] is only valid for these attacks. However, there are ways to extend it to general attacks by using the *postselection technique* mentioned above.

Exercise 5.9 Show that the bound on the secret key rate in (5.69) follows from the formula in (5.64).

Security of BB84

We can also use entropic uncertainty relations to prove the security of the BB84 protocol. Recall that in the BB84 protocol, there are two different measurement bases that Alice and Bob can use: The computational basis (or Z-basis) $\{|0\rangle, |1\rangle\}$ and the Hadamard basis (or X-basis) $\{|+\rangle, |-\rangle\}$. Furthermore, in Sect. 3.3 we have shown that $\log(1/c) = 1$ for the X and Z measurements. The result of Corollary 5.8 then states that

$$H(Z|E) + H(X|B) \geq 1, \tag{5.72}$$

where E denotes Eve's quantum system and B denotes the one that Bob has. The first term in the above relation is evaluated for the state ρ_{ZBE}, which is the post-measurement state after Alice has measured in the Z-basis. Analogously, the second term is evaluated for the post-measurement state after Alice has measured in the X-basis, ρ_{XBE}. With this relation, we can bound Eve's information by

$$H(Z|E) \geq 1 - H(X|B). \tag{5.73}$$

Note that if $H(Z|E) = 1$, then Eve has no knowledge on Alice's state. The fact that measurements cannot decrease the entropy (see Theorem 3.35) implies that $H(Z|E) \geq 1 - H(X|X')$, where X' denotes Bob's measurement in the X-basis.

In the parameter estimation step of the protocol, Alice and Bob estimate the amount of errors in their respective bit strings. Here, two different scenarios can occur:

1. If $\Pr[X \neq X'] \geq 0$, Alice and Bob will detect a certain amount of errors in their sample and abort the protocol.
2. If $\Pr[X = X'] \approx 1$, Alice and Bob can generate a key.

Since in the first case Alice and Bob abort the protocol, let us assume that $\Pr[X \neq X'] = \delta$ with $\delta \approx 0$. With this, we can find an upper bound on $H(X|X')$ terms of δ: Recall that $H(X|X') = H(X, X') - H(X') \leq H(X, X')$. Furthermore, note that the joint probability distribution of X and X' is given by

$$p_{X,X'}(x, x') = \begin{cases} \delta & \text{if } x \neq x' \\ 1 - \delta & \text{if } x = x', \end{cases} \tag{5.74}$$

which yields $H(X, X') = -\delta \log \delta - (1 - \delta) \log(1 - \delta) = h_2(\delta)$, where $h_2(\delta)$ is the binary entropy function. Also note that $H(Z) = 1$ since the outcomes are uniformly distributed. With these results, we can bound the mutual information $I(Z : E)$ (defined in Definition 3.37) between Alice's and Eve's respective systems:

$$I(Z : E) = H(Z) - H(Z|E) \tag{5.75}$$

$$\leq 1 - (1 - H(X|X')) \tag{5.76}$$

$$\leq H(X, X') \tag{5.77}$$

$$\leq h(\delta). \tag{5.78}$$

We now use an identity stated in Corollary 3.43, namely

$$I(Z : E) = D(\rho_{ZE} \| \rho_Z \otimes \rho_E). \tag{5.79}$$

For $\delta \to 0$, $h(\delta) = 0$ (see, e.g., the plot in Fig. 3.2) and therefore $I(Z : E) = 0$. From (5.79) we can then conclude that $D(\rho_{ZE} \| \rho_Z \otimes \rho_E) = 0$ for $\delta \to 0$. Since the relative entropy is zero if and only if the two states are equal (see Theorem 3.42), it follows that $\rho_{ZE} = \rho_Z \otimes \rho_E$. Hence, Eve has no information on Alice's system. To complete the security proof, one has to do the same calculation in the X-basis using the uncertainty relation $H(X|E) + H(Z|B) \geq 1$, but since this is completely analogous to the above derivation, we omit it here.

Using the formula given in (5.71) we can conclude that the asymptotic secure key rate of the BB84 protocol is given by

$$r_\infty^{BB84} \geq 1 - h(e_Z) - h(e_X), \tag{5.80}$$

where e_Z and e_X are the error rates in the Z- and X-bases, respectively.

It is important to note that the above proof only holds when there are no correlations between the individual rounds of the protocol, i.e., it only applies in the case of individual attacks. However, since in general

$$\rho_{A_1 B_1 E_1} \otimes \rho_{A_2 B_2 E_2} \otimes \cdots \otimes \rho_{A_n B_n E_n} \neq \rho_{A_1 A_2 \ldots A_n B_1 B_2 \ldots B_n E_1 E_2 \ldots E_n}, \tag{5.81}$$

this does not imply security against general attacks. This problem can be solved by employing the so-called *de Finetti theorem*, which states that the proof for collective attacks also implies security for general attacks. Applying the de Finetti theorem requires that the states of the protocol are invariant under permutation of the rounds. While this seems like a strong requirement, it can be argued that this is always fulfilled when we assume that the eavesdropper has unlimited power. A rigorous treatment of this issue can be found in [26] and [9].

5.6 Finite-Key Analysis

The analysis of finite-key statistics is, in general, more difficult than the analysis of the infinite-key analysis since the Devetak–Winter rate, which provides a lower bound on the secure key rate, is only valid in the limit of infinitely many rounds of the protocol. However, for practical applications it is inevitable to prove the security of a QKD protocol for keys of finite size since it is usually very costly and time-consuming to generate large amounts of secure key bits.

Recall that the main challenge in modern security proofs of QKD protocols in the infinite scenario lies in bounding the conditional entropy between Alice and Eve, i.e., the information that Eve has on the key. To understand where the difficulties are, let us recap a typical QKD protocol in the finite scenario. In a general QKD protocol, the goal is to output an identical pair of keys $k_A = k_B$, which is completely unknown to Eve. It is also possible that the protocol aborts, which means setting $k_A = k_B = \perp$, but it never outputs an insecure key. The steps of such a protocol in the entanglement-based setting are as follows:

1. The first step of the protocol is to distribute M quantum states such that the joint state of Alice and Bob is ρ_{AB}^M and Eve holds a purification of this state (to account for the worst case scenario).[13]
2. Alice and Bob perform local measurements on their respective states and collect the classical outputs. The allowed sets of measurements depend on the protocol that is chosen. Typically, Alice and Bob choose some round to be used for key generation and others as test rounds.
3. Now the classical post-processing begins. The first step here is parameter estimation, where Alice and Bob estimate the noise in the quantum channel and, therefore, the information that Eve has. For this purpose, they reveal the setting and outcomes of the test rounds as well as a random sample of the key-generation rounds. At the end of this step they hold a pair of partially secret, partially correlated bit strings of length $n < M$ (the raw key), which we denote K_A^n and K_B^n, respectively.
4. The next step in the classical post-processing procedure is error correction. The goal of this step is for Bob to compute a guess $\hat{K}_A^n$ of Alice's raw key K_A^n, given his raw key K_B^n and the information he receives during the error correction procedure. This leaks leak_{EC} bits of information to Eve. Furthermore, Alice and Bob need to check whether the error correction procedure was successful. For this purpose, Alice computes a hash h_A of length $\lceil \log(1/\epsilon_{cor}) \rceil$ from her raw key K_A^n to apply a randomly chosen two-universal hash function.[14] Bob compares this to a hash h_B he computes from his guess $\hat{K}_A^n$. They compare their respective hashes and if $h_A \neq h_B$, they abort the protocol setting $k_A = k_B = \perp$. The total

---

[13]Note that $\rho_{AB}^M \neq \rho_{AB}^{\otimes M}$, i.e., the individual signals are not identical, in order to account for the possibility of coherent attacks.

[14]See also Sect. 4.2.2, where we treated this step in more detail.

amount of information leaked during the error correction process is

$$\text{leak}_{\text{EC}} + \left\lceil \log \frac{1}{\epsilon_{\text{cor}}} \right\rceil \leq \text{leak}_{\text{EC}} + \log \frac{2}{\epsilon_{\text{cor}}}. \tag{5.82}$$

5. In the last step, privacy amplification, Alice and Bob turn their strings into secure keys using a randomness extractor. Alice randomly picks another two-universal hash function and communicates it to Bob. They apply the function to K_A^n and $\hat{K}_A^n$, respectively, and obtain a set of shorter secret keys k_A and k_B of length $l < n$. The length l of the secret key is chosen such that

$$l \leq H_{\min}^\epsilon(K_A^n | E) - \text{leak}_{\text{EC}} - \log \frac{2}{\epsilon_{\text{cor}}} - 2 \log \frac{1}{2\epsilon_{\text{sec}}}, \tag{5.83}$$

for some $\epsilon, \epsilon_{\text{cor}}, \epsilon_{\text{sec}} \geq 0$ as explained in Sect. 4.2.3.

The key rate in the non-asymptotic scenario is then given by

$$r = \frac{\text{length of secret key}}{\text{number of signals exchanged}} = \frac{l}{M}. \tag{5.84}$$

Exercise 5.10 Show that $\lceil \log(1/x) \rceil \leq \log(2/x)$ for all x.

Security Proof
To prove the security of the above protocol in the case of a finite key we employ Theorem 5.4 that allows us to treat correctness and secrecy of the protocol individually.

The correctness of the key is established in the error correction step. Recall from Sect. 4.2.2 that when Alice and Bob use a two-universal hash function f_{EC} to check whether the error correction procedure has been successful, the probability that their bit strings differ even though the hashes are equal is given by

$$\Pr[K_A^n \neq K_B^n | f_{\text{EC}}(K_A^n) = f_{\text{EC}}(K_B^n)] \leq \epsilon_{\text{cor}}. \tag{5.85}$$

Furthermore, note that the privacy application procedure does not affect the correctness of the protocol (see Sect. 4.2.3): let $k_A = f_{\text{PA}}(K_A^n)$ be Alice's key after she has applied the two-universal hash function f_{PA} in the privacy amplification step (and similar for Bob's key). The probability that the keys differ is then given by

$$\Pr[k_A \neq K_B] \leq \Pr[K_A^n \neq K_B^n] \leq \epsilon_{\text{cor}}. \tag{5.86}$$

Note that if the hashes differ and the protocol aborts, the keys are trivially equal since their value is set to $\perp$. Hence, we have

$$\Pr[k_A \neq k_B] \leq \epsilon_{\text{cor}}. \tag{5.87}$$

To prove the secrecy of the protocol we use the quantum leftover hash lemma (see Lemma 4.9) and a chain rule for the conditional min-entropy: First, recall that the quantum leftover hash lemma tells us that after applying a two-universal hash function in the privacy amplification step it holds that

$$\frac{1}{2}\|\rho_{k_A E_{tot}} - \rho_U \otimes \rho_{E_{tot}}\| \leq 2\epsilon' + \frac{1}{2}\sqrt{2^{l-H_{min}^{\epsilon'}(K_A^n|E_{tot})}}, \qquad (5.88)$$

where E_{tot} includes all the information that is available to Eve: the purifying system E, the classical information C that has been publicly exchanged during the error correction step as well as the knowledge Y of the two-universal hash function used in the privacy amplification step. We now use the following chain rule for the min-entropy:

$$H_{min}^{\epsilon'}(K_A^n|CE) \geq H_{min}^{\epsilon'}(K_A^n|E) - \log|C|, \qquad (5.89)$$

where $\log|C| = \text{leak}_{EC} + \log\frac{2}{\epsilon_{EC}}$ (see the discussion of the error correction step above). In summary, we can now show that

$$\frac{1}{2}\|\rho_{k_A E_{tot}} - \rho_U \otimes \rho_{E_{tot}}\| \leq 2\epsilon' + \frac{1}{2}\sqrt{2^{l-H_{min}^{\epsilon'}(K_A^n|E)-\text{leak}_{EC}-\log\frac{2}{\epsilon_{EC}}}} \qquad (5.90)$$

$$\leq 2\epsilon' + \frac{1}{2}\sqrt{2^{\log(2\epsilon_{sec})^2}} \qquad (5.91)$$

$$\leq 2\epsilon' + \epsilon_{sec}, \qquad (5.92)$$

where in the second inequality we have used the expression for the key length (5.83). Combining the two parts of the proof, we find that the protocol described above is ϵ-secure with $\epsilon = \epsilon_{cor} + 2\epsilon' + \epsilon_{sec}$.

Finding a bound on the conditional min-entropy on the right hand side of (5.83) is arguably the most difficult part of a security proof since it cannot be computed directly due to the fact that Alice and Bob do not have access to Eve's system E. Hence, the main challenge for any security proof is to bound the conditional min-entropy. In the next section, we discuss how one can use entropic uncertainty relations to achieve this task by studying (once again) the example of the BB84 protocol.

Finite-Key Analysis of BB84

In the finite-key regime, we can also make use of entropic uncertainty relations to bound the conditional min-entropy. We demonstrate this with the example of the BB84 protocol as it was presented in [31]. To analyse the security of a protocol in the finite-key regime, we can employ the entropic uncertainty relation for smooth one-shot entropies that has been presented in Theorem 3.55. Slightly reformulating the uncertainty relation from Theorem 3.55 yields

$$H^{\epsilon}_{\min}(K^Z_A|E) + H^{\epsilon}_{\max}(K^X_A|K^X_B) \geq n \log \frac{1}{c}, \tag{5.93}$$

where K^Z_A is the key string that Alice gets from measuring in the Z-basis (K^X_A and K^X_B are defined analogously). Note that for the measurements that appear in the BB84 protocol, the X and Z measurements, we have that $\log(1/c) = 1$. Using this and slightly rewriting the above expression yield

$$H^{\epsilon}_{\min}(K^Z_A|E) \geq n - H^{\epsilon}_{\max}(K^X_A|K^X_B), \tag{5.94}$$

which shows that we can bound the information that an eavesdropper has on the key generated by measuring in the Z-basis by finding an upper bound on the max-entropy.

Recall that from the parameter estimation step described in Sect. 4.2.1 we get a probabilistic bound on the number of errors: the probability that the error rate in the string used for key generation, Λ_n, is larger than the error rate observed in the sample, Λ_k, plus a small constant γ, given that the sample error rate is below a certain threshold $\lambda_{\max}$, is given by

$$\Pr[\Lambda_n \geq \Lambda_k + \gamma | \Lambda_k \leq \lambda_{\max}] \leq \frac{e^{-\frac{2k^2 n \gamma^2}{(k+1)N}}}{p_{\text{pass}}}, \tag{5.95}$$

where k is the size of the sample, n is the number of bit strings that are left after the parameter estimation step, $N = n + k$, and p_{pass} is the probability that the error rate in the sample size is lower than the threshold $\lambda_{\max}$. We can use this probabilistic bound to derive an upper bound on the max-entropy.

For this purpose, we first define two probability distributions: The first one is the joint probability distribution of Alice's and Bob's respective keys K_A and K_B conditioned on the sample passing the test:

$$P_{K_A K_B \Lambda_k}(k_A, k_B, \lambda_k) = \Pr[K_A = k_A, K_B = k_B, \lambda_k = \lambda_k | \Lambda_k \leq \lambda_{\max}]. \tag{5.96}$$

The second probability distribution we consider is the following:

$$Q_{K_A K_B \Lambda_k}(k_A, k_B, \lambda_k) = \begin{cases} \frac{P_{K_A K_B \Lambda_k}(k_A, k_B, \lambda_k)}{\Pr[\Lambda_n < \Lambda_k + \gamma | \Lambda_k \leq \lambda_{\max}]}, & \text{if } \lambda_n < \lambda_k + \gamma \\ 0 & \text{otherwise.} \end{cases} \tag{5.97}$$

We need the second distribution because in this case we know that the error rate in the remaining bits fulfils $\Lambda_n < \Lambda_k + \gamma \leq \lambda_{\max} + \gamma$ with probability 1. As a result, the number of errors on the remaining n bits, defined as $W = n\Lambda_n$, satisfies

$$W \leq \lfloor n(\lambda_{\max} + \gamma) \rfloor. \tag{5.98}$$

In order to find a bound on the max-entropy using probability distribution Q we need that Q and P are close to each other with respect to the purified distance, which follows directly from the definition of the max-entropy. To see this, note that the fidelity of P and Q is given by

$$F(P, Q) = \left(\sum_{k_A, k_B, \lambda_k} \sqrt{P_{K_A K_B \Lambda_k}(k_A, k_B, \lambda_k) \, Q_{K_A K_B \Lambda_k}(k_A, k_B, \lambda_k)} \right)^2 \tag{5.99}$$

$$= \left(\sum_{k_A, k_B, \lambda_k, \lambda_n \leq \lambda_k + \gamma} \frac{P_{K_A K_B \Lambda_k}(k_A, k_B, \lambda_k)}{\sqrt{\Pr[\Lambda_n \geq \Lambda_k + \gamma | \Lambda_k \leq \lambda_{\max}]}} \right)^2 \tag{5.100}$$

$$= \Pr[\Lambda_n < \Lambda_k + \gamma | \Lambda_k \leq \lambda_{\max}] \tag{5.101}$$

$$\geq 1 - \frac{e^{-\frac{2k^2 n \gamma^2}{(k+1)N}}}{p_{\text{pass}}}. \tag{5.102}$$

From the definition of the purified distance it now directly follows that

$$P(P, Q) \leq \sqrt{\frac{e^{-\frac{2k^2 n \gamma^2}{(k+1)N}}}{p_{\text{pass}}}} = \frac{e^{-\frac{k^2 n \gamma^2}{(k+1)N}}}{\sqrt{p_{\text{pass}}}}, \tag{5.103}$$

and hence, P and Q are ϵ-close with respect to the purified distance for $\epsilon = \frac{\epsilon'}{\sqrt{p_{\text{pass}}}}$ and $\gamma = \sqrt{\frac{(k+1)N}{k^2 n} \ln \frac{1}{\epsilon'}}$.

This helps us to bound the max-entropy: since the distribution P (which is the actual distribution we are interested in) and the distribution Q are ϵ-close, from the definition of the max-entropy it follows that

$$H_{\max}^{\epsilon}(K_A | K_B)_P \leq H_{\max}(K_A | K_B)_Q. \tag{5.104}$$

The expression on the right hand side can be further bounded by using the definition of the max-entropy for the classical case, which is

$$H_{\max}(X | Y)_P = \max_{y \in \mathcal{Y}} \log |\operatorname{supp} P_{X | Y = y}|, \tag{5.105}$$

where $\mathcal{Y}$ is the outcome set that corresponds to the probability distribution P_Y. Since the probability distribution Q has only support on strings that fulfil $\lambda_n < \lambda_k + \gamma \leq \lambda_{\max} + \gamma$, we only need to count how many strings of length n have at most $\lambda_{\max} + \gamma$

errors. This yields

$$H_{\max}\left(K_A|K_B\right)_Q \leq \log \sum_{w=0}^{\lfloor n(\lambda_{\max}+\gamma)\rfloor} \binom{n}{w} \leq nh_2(\lambda_{\max} + \gamma), \qquad (5.106)$$

where the last inequality was shown in [32] with h_2 being the binary entropy. Altogether, we can now bound the information an eavesdropper has on the outcomes of the Z measurement by

$$H_{\min}^\epsilon\left(K_A^Z|E\right) \geq n(1 - h_2(\lambda_{\max} + \gamma)). \qquad (5.107)$$

To arrive at the secure key, Alice and Bob now perform error correction and privacy amplification, which leaks additional information about the random variable Z. As a result, the length l of the final key is given by

$$l \geq n(1 - h_2(\lambda_{\max} + \gamma)) - \text{leak}_{\text{EC}} - \log \frac{1}{\epsilon_{\text{cor}}} - 2\log \frac{1}{2\epsilon_{\text{sec}}}. \qquad (5.108)$$

Note that by bounding the smooth min-entropy Alice and Bob *almost* fulfil the conditions needed to apply a randomness extractor to Z in order to generate the final secure key (the privacy amplification step). From Definition 4.8 it directly follows that if $H_{\min}(Z|E) > k$, applying a (k, ϵ)-strong quantum-proof randomness extractor yields an ϵ-secure key in the sense of Definition 5.3. However, so far we have only established a bound on the *smooth* min-entropy and not on the min-entropy itself. Fortunately, we can show that if $H_{\min}^{\tilde{\epsilon}}(Z|E) \geq k$, a (k, ϵ)-strong quantum-proof extractor will give a key that is $(\epsilon + 2\tilde{\epsilon})$-secure.

Recall that by definition, $H_{\min}^{\tilde{\epsilon}}(Z|E)$ is the min-entropy that maximizes $H_{\min}(Z|E)$ in an $\tilde{\epsilon}$-ball around the state ρ_{ZE}. Suppose that $\tilde{\rho}_{ZE} \in \mathcal{B}^{\tilde{\epsilon}}(\rho_{ZE})$ is the state that maximizes the min-entropy. Then

$$H_{\min}(Z|E)_{\tilde{\rho}} = H_{\min}^{\tilde{\epsilon}}(Z|E)_\rho \geq k. \qquad (5.109)$$

Hence, by the definition of a (k, ϵ)-strong quantum-proof randomness extractor we have that

$$D\left(\tilde{\rho}_{\text{Ext}(Z,Y)YE}, \rho_U \otimes \rho_Y \otimes \rho_E\right) \leq \epsilon. \qquad (5.110)$$

We can now use the triangle inequality for the trace distance twice in order to relate the actual distance we are interested in to quantities where we can find an upper bound:

$$D\left(\rho_{\text{Ext}(Z,Y)YE}, \rho_U \otimes \rho_Y \otimes \rho_E\right) \leq D\left(\rho_{\text{Ext}(Z,Y)YE}, \tilde{\rho}_{\text{Ext}(Z,Y)YE}\right) \qquad (5.111)$$

$$+ D\left(\tilde{\rho}_{\text{Ext}(Z,Y)YE}, \rho_U \otimes \rho_Y \otimes \tilde{\rho}_E\right) + D\left(\rho_U \otimes \rho_Y \otimes \tilde{\rho}_E, \rho_U \otimes \rho_Y \otimes \rho_E\right). \qquad (5.112)$$

The first element in the sum on the right hand side can be bounded by using the fact that applying the extractor function Ext does not increase the trace distance and, furthermore, that system Y is the same for both ρ and $\tilde{\rho}$:

$$D\left(\rho_{\mathsf{Ext}(Z,Y)YE}, \tilde{\rho}_{\mathsf{Ext}(Z,Y)YE}\right) \leq D\left(\rho_{ZE} \otimes \rho_Y, \tilde{\rho}_{ZE} \otimes \rho_Y\right) \qquad (5.113)$$

$$\leq P\left(\rho_{ZE}, \tilde{\rho}_Z\right) = \tilde{\epsilon}. \qquad (5.114)$$

In the last step we have exploited the fact that the purified distance is always greater or equal to the trace distance, which follows directly from Theorem 2.59. This property is also used to bound the third element in the sum:

$$D\left(\rho_U \otimes \rho_Y \otimes \tilde{\rho}_E, \rho_U \otimes \rho_Y \otimes \rho_E\right) \leq P\left((\tilde{\rho}_{ZE}, \rho_{ZE}\right) = \tilde{\epsilon}. \qquad (5.115)$$

Altogether, we find that

$$D\left(\rho_{\mathsf{Ext}(Z,Y)YE}, \rho_U \otimes \rho_Y \otimes \rho_E\right) \leq \epsilon + 2\tilde{\epsilon}, \qquad (5.116)$$

which shows that by bounding the smooth min-entropy, we can still find a quantum-proof randomness extractor that gives us a secret key.

References

1. Beaudry, N.J.: Assumptions in quantum cryptography. Ph.D. thesis, ETH Zurich (2014). https://doi.org/10.3929/ETHZ-A-010432410
2. Bechmann-Pasquinucci, H., Gisin, N.: Incoherent and coherent eavesdropping in the six-state protocol of quantum cryptography. Phys. Rev. A **59**(6), 4238–4248 (1999). https://doi.org/10.1103/physreva.59.4238
3. Ben-Or, M., Horodecki, M., Leung, D.W., Mayers, D., Oppenheim, J.: The universal composable security of quantum key distribution. In: Theory of Cryptography, pp. 386–406. Springer, Berlin, Heidelberg (2005). https://doi.org/10.1007/978-3-540-30576-7_21
4. Bennett, C.H., Brassard, G.: Quantum cryptography: public key distribution and coin tossing. Proc. IEEE Int. Conf. Comput. Syst. Signal Process. **175**, 8 (1984)
5. Berta, M., Christandl, M., Colbeck, R., Renes, J.M., Renner, R.: The uncertainty principle in the presence of quantum memory. Nat. Phys. **6**(9), 659–662 (2010). https://doi.org/10.1038/nphys1734
6. Biham, E., Boyer, M., Boykin, P.O., Mor, T., Roychowdhury, V.: A proof of the security of quantum key distribution. J. Cryptol. **19**(4), 381–439 (2006). https://doi.org/10.1007/s00145-005-0011-3. First appeared in a shorter version in Proceedings of the Thirty-Second Annual ACM Symposium on Theory of Computing - STOC '00
7. Bruß, D.: Optimal eavesdropping in quantum cryptography with six states. Phys. Rev. Lett. **81**(14), 3018–3021 (1998). https://doi.org/10.1103/physrevlett.81.3018
8. Bruß, D., Meyer, T.: Quantum cryptography. In: Quantum Information, Computation and Cryptography, pp. 277–308. Springer, Berlin, Heidelberg (2010). https://doi.org/10.1007/978-3-642-11914-9_9
9. Christandl, M., König, R., Renner, R.: Postselection technique for quantum channels with applications to quantum cryptography. Phys. Rev. Lett. **102**(2) (2009). https://doi.org/10.1103/physrevlett.102.020504

10. Cirac, J.I., Gisin, N.: Coherent eavesdropping strategies for the four state quantum cryptography protocol. Phys. Lett. A **229**(1), 1–7 (1997). https://doi.org/10.1016/s0375-9601(97)00176-x

11. Colbeck, R., Renner, R.: No extension of quantum theory can have improved predictive power. Nat. Commun. **2**(1) (2011). https://doi.org/10.1038/ncomms1416

12. Csiszar, I., Korner, J.: Broadcast channels with confidential messages. IEEE Trans. Inform. Theory **24**(3), 339–348 (1978). https://doi.org/10.1109/tit.1978.1055892

13. Devetak, I., Winter, A.: Distillation of secret key and entanglement from quantum states. Proc. R. Soc. A: Math. Phys. Eng. Sci. **461**(2053), 207–235 (2005). https://doi.org/10.1098/rspa.2004.1372

14. Fuchs, C.A., Gisin, N., Griffiths, R.B., Niu, C.S., Peres, A.: Optimal eavesdropping in quantum cryptography. I. Information bound and optimal strategy. Phys. Rev. A **56**(2), 1163–1172 (1997). https://doi.org/10.1103/physreva.56.1163

15. Grasselli, F.: Quantum Cryptography. Springer International Publishing, Cham (2021). https://doi.org/10.1007/978-3-030-64360-7

16. Hwang, W.Y.: Quantum key distribution with high loss: toward global secure communication. Phys. Rev. Lett. **91**(5) (2003). https://doi.org/10.1103/physrevlett.91.057901

17. König, R., Renner, R., Bariska, A., Maurer, U.: Small accessible quantum information does not imply security. Phys. Rev. Lett. **98**(14) (2007). https://doi.org/10.1103/physrevlett.98.140502

18. Lo, H.K., Chau, H.F.: Unconditional security of quantum key distribution over arbitrarily long distances. Science **283**(5410), 2050–2056 (1999). https://doi.org/10.1126/science.283.5410.2050

19. Lo, H.K., Ma, X., Chen, K.: Decoy state quantum key distribution. Phys. Rev. Lett. **94**(23) (2005). https://doi.org/10.1103/physrevlett.94.230504

20. Maurer, U.: Constructive cryptography - a new paradigm for security definitions and proofs. In: Theory of Security and Applications, pp. 33–56. Springer, Berlin, Heidelberg (2012). https://doi.org/10.1007/978-3-642-27375-9_3

21. Maurer, U., Renner, R.: Abstract cryptography. In: Proceedings of Innovations in Computer Science, ICS 2010, pp. 1–21 (2011)

22. Mayers, D.: Unconditional security in quantum cryptography. J. ACM **48**(3), 351–406 (2001). https://doi.org/10.1145/382780.382781

23. Nielsen, M.A., Chuang, I.L.: Quantum Computation and Quantum Information. Cambridge University Press, Cambridge (2000)

24. Portmann, C., Renner, R.: Security in Quantum Cryptography (2021) arXiv preprint: https://arxiv.org/abs/1409.3525

25. Renes, J.M., Boileau, J.C.: Conjectured strong complementary information tradeoff. Phys. Rev. Lett. **103**(2) (2009). https://doi.org/10.1103/physrevlett.103.020402

26. Renner, R.: Security of quantum key distribution. Ph.D. thesis, ETH Zurich (2005). https://doi.org/10.3929/ethz-a-005115027

27. Renner, R.: Quantum Information Theory (Lecture notes) (2013). http://edu.itp.phys.ethz.ch/hs15/QIT/renner_lecture_notes12.pdf

28. Renner, R., König, R.: Universally composable privacy amplification against quantum adversaries. In: Theory of Cryptography, pp. 407–425. Springer, Berlin, Heidelberg (2005). https://doi.org/10.1007/978-3-540-30576-7_22

29. Scarani, V., Kurtsiefer, C.: The black paper of quantum cryptography: real implementation problems. Theoret. Comput. Sci. **560**, 27–32 (2014). https://doi.org/10.1016/j.tcs.2014.09.015

30. Shor, P.W., Preskill, J.: Simple proof of security of the BB84 quantum key distribution protocol. Phys. Rev. Lett. **85**(2), 441–444 (2000). https://doi.org/10.1103/physrevlett.85.441

31. Tomamichel, M., Lim, C.C.W., Gisin, N., Renner, R.: Tight finite-key analysis for quantum cryptography. Nat. Commun. **3**(1) (2012). https://doi.org/10.1038/ncomms1631

32. van Lint, J.H.: Introduction to Coding Theory. Springer, Berlin, Heidelberg (1982). https://doi.org/10.1007/978-3-662-07998-0

33. Wang, X.B.: Beating the photon-number-splitting attack in practical quantum cryptography. Phys. Rev. Lett. **94**(23) (2005). https://doi.org/10.1103/physrevlett.94.230503
34. Wilde, M.M.: Quantum Information Theory, 2nd edn. Cambridge University Press, Cambridge (2017). https://doi.org/10.1017/9781316809976

Device-Independent QKD

6

Abstract

Device-Independent Quantum Key Distribution (DIQKD) deals with the problem that security proofs often only apply to the ideal setting of the theoretical protocol and do not take into account the problems that arise when implementing a protocol, such as the photon-number-splitting attack. DIQKD circumvents this problem by assuming that the incorporated devices are not trusted, therefore not making any assumptions on them in the security proof. To prove the security of the protocol, Alice and Bob instead need to verify that the input-output statistics of the devices violates a Bell inequality.

6.1 Device-Independent Concepts

The analysis of device-independent quantum processing tasks requires different techniques than the device-dependent analysis, since in the device-independent scenario we cannot make any assumptions on the states and devices that occur. In this section, we introduce the concepts needed to understand how secure key generation in the device-independent setting works. A detailed treatment of device-independent quantum information processing can, for example, be found in [3].

6.1.1 Black Boxes

Since in DIQKD we do not make any assumptions on the devices that are involved, we need to treat them differently than in the device-dependent case. Our goal is to use certain properties of the physical devices without having to make statements about their internal workings. Hence, instead of characterizing a device by its hardware we treat it as a *black box* with some buttons. The user can press these

© The Author(s), under exclusive license to Springer Nature Switzerland AG 2021
R. Wolf, *Quantum Key Distribution*, Lecture Notes in Physics 988,
https://doi.org/10.1007/978-3-030-73991-1_6

Fig. 6.1 Black boxes. Without making any assumptions about the inner workings of the boxes, Alice and Bob can press a button (or one of several buttons) of the box, which represents the input x/y of the box. They will get some output a/b. This yields the input-output statistics of the boxes

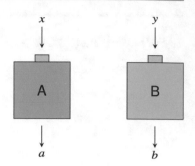

buttons and get some output. The only information the user of the box can then deduce is the input-output statistics of the box.

The input-output behaviour of the black box is described by the conditional probability distribution $p_{O|I}$, where O is the random variable that represents the possible output of the box and I is the corresponding random variable for the possible inputs. For instance, if the user can press four different buttons, then I is a random variable over the set of possible inputs $\{0, 1, 2, 3\}$. Analogously, if the output of the box is a classical bit, then $O = \{0, 1\}$. The probability distribution $p_{O|I}$ then tells you the probability of each of the possible outputs conditioned on the input button you have pressed. For example, if the probability distribution is uniform when you press the 0-button, then

$$p_{O|I}(0|0) = \frac{1}{2}, \quad p_{O|I}(1|0) = \frac{1}{2}. \tag{6.1}$$

Consider the case where Alice and Bob each have access to one part of a bipartite black box (see Fig. 6.1), which is the kind of box we are interested with regard to QKD protocols. Crucially, the two parts of the bipartite box are separated in space, which means that Alice and Bob each have access to their part of the box but cannot access the other party's box. Hence, Alice can input something to her part of the box and read the corresponding output, but she has no access to Bob's component (and the other way round).

Mathematically, the boxes are fully described by the conditional probability distribution $p_{AB|XY}$, where X and A are the random variables describing Alice's input and output, respectively, and Y and B are the random variables describing Bob's input and output, respectively. In general, there are no restrictions on the possible probability distributions, i.e., $p_{AB|XY}$ can be any conditional probability distribution. However, since we want to employ these boxes in the context of quantum key distribution, we assume that the correlations that the boxes exhibit can be explained within the framework of quantum mechanics, hence we call these boxes *quantum boxes*:

Definition 6.1 A quantum box is a bipartite box described by a probability distribution $p_{AB|XY}$ such that there exists a bipartite quantum state ρ_{AB} and two sets of POVMs for Alice and Bob $\{M_a^x\}_{a \in \mathcal{A}}$ for all $x \in \mathcal{X}$ and $\{M_b^y\}_{b \in \mathcal{B}}$ for all $y \in \mathcal{Y}$, respectively, for which

$$p_{AB|XY}(a, b|x, y) = \text{Tr}\left(M_a^x \otimes M_b^y \rho_{AB}\right) \tag{6.2}$$

for all a, b, x, y.

Note that even though we assume that the bipartite box is quantum, we do not make any assumptions on the internal workings of the box. The assumption that there is a bipartite quantum state $\rho_{AB} \in \mathcal{B}(\mathcal{H}_A \otimes \mathcal{H}_B)$ that gives rise to the observed statistics is no limitation since we do not restrict the dimensions of the Hilbert spaces.

If the conditional probability distribution that characterizes the boxes can be explained by classical correlations (i.e., shared randomness instead of a shared quantum state) alone, then the box is classical:

Definition 6.2 A classical box is a bipartite box with conditional probability distribution that can be written in the form

$$p_{AB|XY}(a, b|x, y) = \int_\Lambda d\lambda \, \Pr[\Lambda = \lambda] p_{A|X\Lambda}(a|x, \lambda) p_{B|Y\Lambda}(b|y, \lambda), \tag{6.3}$$

where Λ is the random variable that describes the randomness that is shared by the two parts of the box.

The set of all classical boxes is a subset of the set of all quantum boxes. On the other hand, it is possible to define a type of box that is even more general than quantum boxes, namely *non-signalling boxes*. Here, the only requirement is that the individual components of the boxes produce an output independently of the other component. For instance, if Alice presses a button on her part of the box it will produce an output independent of whether Bob has pressed a button on his part of the box. Mathematically, this means that the marginal probability distributions $p_{A|X}$ and $p_{B|Y}$ are well-defined probability distributions.

Definition 6.3 A non-signalling box is a bipartite box with conditional probability distribution $p_{AB|XY}$ that fulfils the non-signalling conditions

$$\sum_b p_{AB|XY}(a, b|x, y) = \sum_b p_{AB|XY}(a, b|x, y') \tag{6.4}$$

$$\sum_a p_{AB|XY}(a, b|x, y) = \sum_a p_{AB|XY}(a, b|x', y) \tag{6.5}$$

for all a, b and x, x', y, y'.

Fig. 6.2 The different sets of black boxes. The sets of classical, quantum, and non-signalling boxes denote C, Q, and NS, respectively. Note that the inclusions $C \subset Q \subset NS$ are strict. Bell inequalities can be used to separate classical and quantum boxes

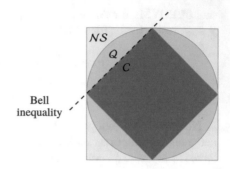

In particular, the no-signalling conditions ensure that in case Alice and Bob's boxes are space-like separated, they cannot be used for instantaneous signalling, which prevents a direct conflict with relativity.

When using these boxes in applications in quantum key distribution, we are interested in distinguishing classical from quantum correlations solely by analysing the input-output statistics of the boxes. This can be achieved by employing *Bell inequalities*. We have seen an example of this when introducing the CHSH inequality in the Ekert protocol, which is a widely-used example of a Bell inequality. In general, there are many such inequalities which separate classical from quantum boxes, as illustrated in Fig. 6.2.

6.1.2 Bell Inequalities

To prove the security of a QKD protocol in the device-independent setting, it is crucial that Alice and Bob are able to certify shared entanglement without having to make assumptions on the states and measurements they use. In other words, they have to be able to distinguish between quantum and classical boxes. Hence, they need a device-independent entanglement witness, for example, a Bell inequality [9]. Bell inequalities define hyperplanes that separate the set of classical boxes C from the set of quantum boxes Q (see Fig. 6.2). The condition that classical boxes are on one side of the hyperplane can be written as a condition on the probability distributions of boxes on C:

$$\forall p_{AB|XY} \in C, \quad \sum_{a,b,x,y} s_{xy}^{ab} \, p_{AB|XY}(a,b|x,y) \leq S, \tag{6.6}$$

for all a, b, x, y, where S and s_{xy}^{ab} are constants. Given a probability distribution $p_{AB|XY}(a,b|x,y)$, it is hence easy to check if the box cannot be a classical one: If

$$\sum_{a,b,x,y} s_{xy}^{ab} \, p_{AB|XY}(a,b|x,y) > S, \tag{6.7}$$

then $p_{AB|XY}$ cannot correspond to a classical box.[1] In other words: All classical boxes fulfil (6.6), while *some* quantum boxes violate this inequality. Hence, whenever a Bell inequality is violated, we can be sure that the underlying boxes are non-classical, and therefore suitable to perform device-independent quantum information processing tasks.

We have already seen a prominent example of a Bell inequality, namely the CHSH inequality, named after its inventors Clauser, Horne, Shimony, and Holt [12]. Here, we consider the case where each party has two measurement choices, $x, y \in \{0, 1\}$, and two possible outcomes, $a, b \in \{-1, +1\}$. The CHSH polynomial is then given by

$$S = \langle a_0 b_0 \rangle + \langle a_0 b_1 \rangle + \langle a_1 b_0 \rangle - \langle a_1 b_1 \rangle, \tag{6.8}$$

where $\langle a_x b_y \rangle = \sum_{a,b} ab \, p_{AB|XY}(a, b|x, y)$, and for all probability distributions that correspond to classical boxes C it holds that

$$S \leq 2. \tag{6.9}$$

As shown in Sect. 4.1.2, this inequality can, for example, be violated if the input-output statistics is produced by making certain measurements on a maximally entangled Bell state. In fact, as shown by Tsirelson in [11], all quantum boxes fulfil

$$S \leq 2\sqrt{2}. \tag{6.10}$$

The violation of a Bell inequality was first shown experimentally in 1972 in [20] and later verified in [6]. More recently, the violation of Bell inequalities was shown taking into account possible loopholes in the experimental setups (which we will discuss in the next chapter), see, for example, [22, 23, 40], and [34].

Non-Local Games

Bell inequalities can also be phrased as the so-called non-local games [13]. In a non-local game, as depicted in Fig. 6.3, Alice and Bob each are asked a question by a referee, who chooses the question according to some probability distribution. Each player only knows about her/his question but does not see the question that is asked the other player. Each player gives an answer and the referee either accepts or rejects the answer according to some predetermined rule. In order to win the game, the players can agree on a strategy before the game begins, but no communication is allowed during the game. If the referee accepts the answers the players win the game.

More formally, we can describe such a game as follows: There are sets of possible questions X and possible answers $\mathcal{A}$ for Alice and sets of possible questions $\mathcal{Y}$

---

[1]Note that the opposite is not true: If the sum on the left hand side is $\leq S$, we cannot make any statement about the underlying black box.

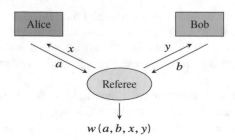

Fig. 6.3 Communication structure of a non-local game. The referee sends questions x and y to Alice and Bob, respectively, and decides according to their answers a and b whether they have won the game ($w(a, b, x, y) = 1$) or lost the game ($w(a, b, x, y) = 0$)

and possible answers $\mathcal{B}$ for Bob. The probability distribution over the questions is denoted q_{XY} and the winning condition for the game is a function $w : \mathcal{A} \times \mathcal{B} \times \mathcal{X} \times \mathcal{Y} \rightarrow \{0, 1\}$. If $w(a, b, x, y) = 1$, the players win the game.

A strategy for the game can be described by a box $p_{AB|XY}$, which is (at least) constraint by the no-signalling condition since Alice and Bob are not allowed to communicate once the game has started. The winning probability for a given box $p_{AB|XY}$ is then[2]

$$\omega(p_{AB|XY}) = \sum_{a,b,x,y} q_{XY}(x, y)p_{AB|XY}(a, b|x, y)w(a, b, x, y). \qquad (6.11)$$

A Bell inequality can now be phrased as such a non-local game in the following way: For classical boxes $p_{AB|XY}$, there is a maximal winning probability $\omega_c < 1$. On the other hand, there exist quantum boxes for which the winning probability is $\omega_q > \omega_c$. Hence, condition (6.6) can be rewritten as

$$\forall p_{AB|XY} \in C, \quad \omega(p_{AB|XY}) \leq \omega_c. \qquad (6.12)$$

Violating a Bell inequality is then equivalent to violating condition (6.12). In other words, if the observed winning probability $\omega(p_{AB|XY})$ of a box $p_{AB|XY}$ is greater than ω_c, it follows that $p_{AB|XY} \notin C$ and the box must be a quantum box. The connection between the winning probability ω of a box and the Bell violation S is given by

$$\omega = \frac{4 + S}{8}. \qquad (6.13)$$

---

[2]Be careful not to confuse the notation for the winning condition, $w(a, b, x, y)$ and the winning probability, $\omega(p_{AB|XY})$.

An explicit example of a non-local game is the CHSH game. Here, Alice and Bob's respective inputs (the questions) and outputs (the answers) are bits, i.e., $a, b, x, y \in \{0, 1\}$. The questions are uniformly distributed, hence $q_{XY}(x, y) = \frac{1}{4}$ for all x, y. The winning condition is given by

$$w_{\text{CHSH}} = \begin{cases} 1 & \text{if } a \oplus b = x \cdot y \\ 0 & \text{otherwise,} \end{cases} \tag{6.14}$$

where $\oplus$ denotes addition modulo 2. The optimal classical strategy for this game yields a winning probability $\omega_c = 0.75$, while the optimal quantum strategy yields a winning probability $\omega_q = \frac{2+\sqrt{2}}{4} \approx 0.85$. The latter is consistent with Tsirelson's bound for the maximal possible CHSH violation for quantum boxes, $S \le 2\sqrt{2}$.

Exercise 6.4 An example for an optimal classical strategy is always choosing the outputs as $(a, b) = (0, 0)$. Show that in this case $\omega_c = 0.75$.

Exercise 6.5 An optimal quantum strategy is given by measuring the maximally entangled state $|\Phi^+\rangle = (|00\rangle + |11\rangle)/\sqrt{2}$ with the following measurements: If $x = 0$, Alice measures the Z-operator and if $x = 1$, she measures the X-operator. For Bob, $y = 0$ corresponds to measuring $(Z + X)/\sqrt{2}$, and $y = 1$ corresponds to measuring $(Z - X)/\sqrt{2}$. Show that this yields a winning probability $\omega_q = \frac{2+\sqrt{2}}{4}$.

6.1.3 Motivation for Device-Independent QKD

To understand why the protocols we have considered so far are not necessarily secure in the device-independent setting, let us study a simple example. Suppose Alice and Bob each have access to one part of a bipartite quantum box (as described above). Both of them have two possible inputs, i.e., $x \in \{0, 1\}$ and $y \in \{0, 1\}$, and two possible outputs, hence $a \in \{0, 1\}$ and $b \in \{0, 1\}$. By performing a number of measurements, Alice and Bob observe the following probability distribution:

$$p(ab|00) = p(ab|11) = \frac{1}{2} \quad \text{if } a = b, \tag{6.15}$$

$$p(ab|01) = p(ab|10) = \frac{1}{4} \quad \text{for all } a, b. \tag{6.16}$$

In other words, whenever Alice and Bob choose the same input ($x = y$), they receive perfectly correlated outcomes. On the other hand, when they choose different inputs ($x \ne y$) the output is completely random (and therefore uncorrelated).

Suppose now that Alice and Bob believe that they have implemented the entanglement-based version of the BB84 protocol as described in Sect. 4.1.2.2. Here, they share a two-qubit state $|\psi\rangle \in \mathbb{C}^2 \otimes \mathbb{C}^2$ and each of them can choose between measuring in the X-basis or in the Z-basis. Let $x, y = 0$ denote a measurement in the X-basis and $x, y = 1$ denote a measurement in the Z-basis. The above probability distribution can then be expressed as follows:

$$\langle \psi | X \otimes X | \psi \rangle = \langle \psi | Z \otimes Z | \psi \rangle = 1, \qquad (6.17)$$

$$\langle \psi | X \otimes Z | \psi \rangle = \langle \psi | Z \otimes X | \psi \rangle = 0, \qquad (6.18)$$

which means that whenever Alice and Bob choose the same measurement basis, they get perfectly correlated results, while measuring in different bases yields completely uncorrelated results. This correlation can only be achieved if the state that Alice and Bob share is the maximally entangled Bell state $|\Phi^+\rangle = (|00\rangle + |11\rangle)/\sqrt{2}$. Hence, Alice and Bob conclude that they share a maximally entangled state, which in turn means that the measurement results can safely be used to produce a secure key.

Exercise 6.6 Show that if Alice and Bob share the state $|\Phi^+\rangle$, they get the correlations listed in (6.17) and (6.18).

However, in the device-independent setting this is no longer true. Here, Alice and Bob can make no assumptions about the Hilbert space dimensions and the measurements they use. The only indication of whether the outputs can be used to generate a secure key is the violation of a Bell inequality. A quick calculation shows that the CHSH value of the statistics given in (6.15) and (6.16) is $S = 0$, which shows that the outputs can *not* be used to generate a secure key! Why is the CHSH inequality not violated, even though we have shown above that the input-output statistics are compatible with Alice and Bob sharing a maximally entangled state?

To arrive at the conclusion that Alice and Bob share a maximally entangled state, we had to make an assumption on the dimension of the Hilbert space of the shared state. However, since it is extremely difficult to verify such an assumption in practice, we would like to avoid having to make it (which is exactly the motivation for device-independent QKD). If we lose the assumption that the Hilbert space is a two-qubit space, we can indeed find a *separable* (and therefore insecure) state that yields the same input-output statistics as the Bell state $|\Phi^+\rangle$. Consider the following bipartite state in the Hilbert space $\mathbb{C}^4 \otimes \mathbb{C}^4$:

$$\rho_{AB} = \frac{1}{4} \sum_{z_0, z_1 \in \{0,1\}} |z_0 z_1\rangle_A \langle z_0 z_1| \otimes |z_0 z_1\rangle_B \langle z_0 z_1|. \qquad (6.19)$$

There is a choice of measurements that exactly produces the probability distribution given in (6.15) and (6.16), namely choosing $x, y = 0$ to be the measurement $Z \otimes \mathbb{I}$ and $x, y = 1$ to be the measurement $\mathbb{I} \otimes Z$. In fact, because the correlation between Alice's and Bob's systems is completely local, it is possible that Eve has a perfect copy of the local state, i.e., that they share the tripartite state

$$\rho_{ABE} = \frac{1}{4} \sum_{z_0, z_1 \in \{0,1\}} |z_0 z_1\rangle_A \langle z_0 z_1| \otimes |z_0 z_1\rangle_B \langle z_0 z_1| \otimes |z_0 z_1\rangle_E \langle z_0 z_1|. \qquad (6.20)$$

This is not possible if the correlations between Alice's and Bob's system are non-local correlation, i.e., those that violate the CHSH inequality. This example shows that in the device-independent setting it is crucial that the observed input-output statistics of the boxes violates a Bell inequality. This ensures that there are non-local correlations, which prevents Eve from having perfect knowledge of the local state.

Exercise 6.7 Show that the input-output statistics given in (6.15) and (6.16) yields a CHSH value of $S = 0$.

Exercise 6.8 Show that using the measurements $0 = Z \otimes \mathbb{I}$ and $1 = \mathbb{I} \otimes Z$ on the state given in (6.19) produces the input-output statistics given in (6.15) and (6.16).

6.2 DIQKD Protocols and Security

The idea of exploiting the violation of a Bell inequality to prove the security of a QKD protocol goes back to the protocol that Ekert proposed in 1991 [19] (see Sect. 4.1.2.1). Later, it was recognized by Mayers and Yao [25] that if the input-output statistics of the devices exhibit a maximal violation of a Bell inequality, the devices can be fully characterized (up to local degrees of freedom) and hence do not need to be trusted. The first security proof of a device-independent QKD protocol was provided by Barrett, Hardy, and Kent [7]. Although the suggested protocol was not useful in practice (it could not tolerate any amount of noise and, furthermore, it only produced a single bit of key during numerous rounds of the protocol), the work showed that secure DIQKD was achievable in principle, which was the beginning of numerous works that explored the application of the device-independent concept in quantum cryptography (see, for example, [2, 26, 30, 32, 45]) and other fields such as randomness expansion and amplification (see, for example, [14, 15, 21, 31, 44]).

6.2.1 Security Against Collective Attacks

We begin our discussion of DIQKD protocols with the protocol described in [2] and [30], which builds on ideas of the Ekert protocol. The general scheme is depicted in Fig. 6.4. The idea of the protocol is the following: A source that is controlled by Eve (assuming the worst case) distributes possibly entangled states to Alice and Bob. Alice and Bob then perform measurements in order to obtain a secret key and to calculate the CHSH violation S. Alice can choose between three measurements A_0, A_1, and A_2, each with possible outcomes $a \in \{-1, +1\}$, while Bob can choose between two measurements B_1 and B_2, also with possible outcomes $b \in \{-1, +1\}$. The raw key (i.e., the bits that are used to generate the secure key by doing error correction and privacy amplification) is extracted from the outcomes of the pair $\{A_0, B_1\}$, which allows us to define the so-called *Quantum Bit Error Rate* (QBER)

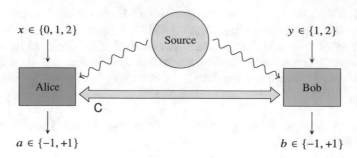

Fig. 6.4 Device-independent QKD protocol. In the DIQKD scenario, a source (usually assumed to be controlled by Eve) distributes states to Alice and Bob. In this protocol, Alice can choose between three different measurements A_x with $x \in \{0, 1, 2\}$ with outcomes $a \in \{-1, +1\}$, while Bob can choose between two different measurements B_y with $y \in \{1, 2\}$, also with outcomes $b \in \{-1, +1\}$. Alice and Bob additionally have access to an authenticated classical communication channel C. Note that in the DIQKD scenario, the measurement devices are untrusted, hence in the worst case they are controlled by Eve

Q as the probability that Alice and Bob get different outcomes when measuring the pair $\{A_0, B_1\}$, i.e.,

$$Q = p_{AB|XY}(a \neq b|01). \tag{6.21}$$

From the QBER, Alice and Bob can estimate the amount of correlations between their outcomes, which in turn yields an estimate of how much classical communication is needed for the error correction process. The measurements A_1, A_2, B_1, and B_2 are used to evaluate the CHSH polynomial (see (6.8)). As a result, Alice and Bob have access to two parameters, the QBER Q and the CHSH violation S, to estimate Eve's information.

Summing up, a general QKD protocol in the device-independent setting can be divided into three phases:

1. **Quantum transmission phase.** Alice and Bob use their devices to perform measurements and generate their respective n-bit strings $\mathbf{a}_0 = a_1 a_2 \ldots a_n$ (the outcomes of Alice measuring A_0) and $\mathbf{b}_1 = b_1 b_2 \ldots b_n$ (the outcomes of Bob measuring B_1).
2. **Parameter estimation phase.** In the second phase, Alice and Bob exchange classical information to estimate the important parameters of the protocol, namely the Bell violation S and the QBER Q. If the parameters allow for the generation of a secure key, i.e., the Bell violation is sufficiently high and the QBER is sufficiently low, they proceed. Otherwise, they abort the protocol.
3. **Classical post-processing phase.** In the last phase, Alice and Bob use the estimate they have on Eve's potential knowledge on the key to perform error correction in order to produce the raw keys and perform privacy amplification to generate the final secure key.

Here, we have marked the parameter estimation step as an extra phase and not as part of the classical post-processing (in contrast to the device-dependent QKD scheme) to emphasize its importance in the DI setting.

A particular implementation of the protocol described above which uses qubits is similar to the Ekert protocol and works as follows: The state ρ_{AB} that is shared by Alice and Bob is the Bell state $|\Phi^+\rangle$ after going through a depolarizing channel (see Exercise 2.30):

$$\rho_{AB} = p|\Phi^+\rangle\langle\Phi^+| + (1+p)\frac{\mathbb{I}}{4} \tag{6.22}$$

and the measurements are given by

$$A_0 = Z, \qquad\qquad B_1 = Z,$$

$$A_1 = \frac{1}{\sqrt{2}}(Z+X), \qquad B_2 = X,$$

$$A_2 = \frac{1}{\sqrt{2}}(Z-X).$$

The resulting correlations then give the following values for the two important parameters, the CHSH violation S and the QBER Q:

$$S = 2\sqrt{2}p \tag{6.23}$$

$$Q = \frac{1}{2} - \frac{p}{2}, \tag{6.24}$$

which in turn yields $S = 2\sqrt{2}(1-2Q)$.[3] It is important to stress that even though these values of S and Q can be achieved by the implementation described above, the security of the protocol does not rely on any of the specifications of the implementation. It only relies on the observed values of S and Q, independent of how they have been generated.

Exercise 6.9 Verify Eqs. (6.23) and (6.24) for the implementation described above.

Security Analysis
The greatest challenge in proving the security of a device-independent protocol is the fact that we cannot make any assumptions on the internal workings of the devices. We only have access to the input-output statistics they produce. One way to prove the security of the protocol described above is finding a bound on the secret key rate r that was introduced in the previous chapter and which is given by the

---

[3]Note that, a priori, there is no relation between the values of S and Q. They are two independent parameters available to estimate Eve's information.

Devetak-Winter rate r_{DW} [16]:[4]

$$r \geq r_{DW} = I(A_0 : B_1) - \chi(B_1 : E), \tag{6.25}$$

where $I(A_0 : B_1) = H(A_0) + H(B_1) - H(A_0, B_1)$ is the mutual information between Alice's measurement A_0 and Bob's measurement B_1, which is the pair of measurements used for key generation. The second quantity,[5] $\chi(B_1 : E)$, is the Holevo quantity given by

$$\chi(B_1 : E) = H(\rho_B) - \frac{1}{2} \sum_{b_1 = \pm 1} H(\rho_{E|b_1}). \tag{6.26}$$

Here, we use the notation $H(\rho)$ to denote the von Neumann entropy in order to distinguish it from the Shannon entropy $H(A)$. The state $\rho_E = \text{Tr}_{AB}(|\psi_{ABE}\rangle\langle\psi_{ABE}|)$ represents the state of Eve's quantum system, and $\rho_{E|b_1}$ denotes the state of Eve's system conditioned on Bob obtaining the result b_1 when measuring B_1. The optimal collective attack for Eve is to prepare the states such that $|\psi_{ABE}\rangle$ is the purification of ρ_{AB}. Note that since the Devetak-Winter rate only holds for collective attacks (and not coherent attacks), this security proof only covers collective attacks.

The first term is straightforward to calculate: Without loss of generality, we assume uniform marginals in the probability distribution, i.e., $\langle a_i \rangle = \langle b_i \rangle = 0$.[6] Then the mutual information between Alice and Bob is given by

$$I(A_0 : B_1) = 1 - h_2(Q), \tag{6.27}$$

where h_2 is the binary entropy.

The second term in the bound is more complicated to estimate. In [30], the authors have found the following expression for the case of uniform marginals:

$$\chi(B_1 : E) \leq h_2\left(\frac{1 + \sqrt{(S/2)^2 - 1}}{2}\right), \tag{6.28}$$

given any violation S of the CHSH inequality. The main idea of the proof for this bound is the fact that it is possible to reduce the problem to a two-qubit optimization problem. The reason for this is that it is possible to decompose any pair of quantum

---

[4]This form of the Devetak-Winter-rate differs slightly from the form we have introduced in the previous chapter, but it directly follows from the Holevo bound: $I(A : B) \leq \chi(A : B)$ [28].

[5]Note that we use $\chi(B_1 : E)$ here and not $\chi(A_0 : E)$ since for the described protocol it holds that $\chi(A_0 : E) \geq \chi(B_1 : E)$ [1]. This means that it is beneficial to use public communication from Bob to Alice in the classical post-processing instead of communication from Alice to Bob.

[6]Note that this is no restriction: If the marginals are not uniform, Alice and Bob could make them so by using public one-way communication and agreeing on randomly flipping a chosen half of their bits. This operation does not change the values of S and Q and is known to Eve.

binary measurements as the direct sum of pairs of measurements which act on two-dimensional spaces.

In summary, the lower bound for the secret key rate in the device-independent scenario is given by

$$r \geq 1 - h_2(Q) - h_2\left(\frac{1 + \sqrt{(S/2)^2 - 1}}{2}\right). \tag{6.29}$$

Exercise 6.10 Show that in case $S \leq 2$ (i.e., in the classical regime) the secret key rate given in (6.29) can never be positive.

To be able to compare the device-independent secret key rate, we can also calculate a device-dependent bound for the implementation of the protocol that we have described above. Here, we now assume that Alice and Bob have perfect control over the states and measurements involved in the protocol. In this case, we have that

$$\chi(B_1 : E) \leq h_2\left(Q + \frac{S}{2\sqrt{2}}\right) \tag{6.30}$$

as shown in [30], while the mutual information between Alice and Bob is the same as in the device-independent case. For the case that $S = 2\sqrt{2}(1 - 2Q)$ the two different key rates are plotted in Fig. 6.5 as a function of the QBER. One can see that the rate is strictly lower in the device-independent setting (as expected), but it is still possible to extract a secret key up to a QBER of $\approx 7.1\%$. The plot also shows the critical QBER of $\approx 11\%$ for the BB84 protocol.

There is an explicit attack which saturates the bound in the device-independent scenario which illustrates the conceptual differences between the device-

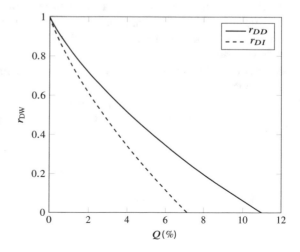

Fig. 6.5 Secret key rate. The plot compares the secret key rate for the Device-Dependent (DD) scenario (solid line) and the Device-Independent (DI) protocol (dashed line). The maximum value for the QBER where a key can still be extracted is $\approx 11\%$ in the DD-scenario and $\approx 7.1\%$ for the DI scenario

independent and the device-dependent setting. The attack works as follows: Eve sends the state

$$\rho_{AB}(S) = \frac{1+C}{2} P_{\Phi^+} + \frac{1-C}{2} P_{\Phi^-}, \tag{6.31}$$

where $C = \sqrt{(S/2)^2 - 1}$ to Alice and Bob. She furthermore defines the measurements that Alice and Bob are allowed to do as

$$A_1 = \frac{1}{\sqrt{1+C^2}} Z + \frac{1}{\sqrt{1+C^2}} X, \quad B_1 = Z, \tag{6.32}$$

$$A_2 = \frac{1}{\sqrt{1+C^2}} Z - \frac{1}{\sqrt{1+C^2}} X, \quad B_2 = X, \tag{6.33}$$

and A_0 is the Z-measurement with probability $1 - 2Q$ and a randomly chosen bit with probability $2Q$. In this way, it is possible to obtain any value of Q. One can now show that for $B_1 = Z$ the Holevo quantity $\chi(B_1 : E)$ is, in fact, *equal* to the right hand side in (6.28), which means that this attack saturates the bound on the secret key rate. In the usual (device-dependent) setting, this attack is not possible since both the state and the measurements depend on S and Q. This violates the assumption that Eve does not have access to Alice's and Bob's respective labs.

Exercise 6.11 What is the probability that Alice and Bob get different outcomes when measuring the pair $\{A_0, B_1\}$, i.e., the value of Q?

Exercise 6.12 Calculate the CHSH violation S for the state and measurements given above.

Finite-Key Analysis
When taking into accounts only a finite amount of protocol rounds (as it is in a realistic description of an implementation), recall that the main task is to bound the min-entropy, which determines the length of the secret key (see (5.83)).

If we limit the eavesdropper to collective attacks we can work under the IID assumption (IID = independent and identically distributed), which allows us to use some theorems that simplify the security analysis. In the IID scenario, each round of the protocol is independent of the other rounds and all rounds are identical. This means that the state ρ_{AB}^M of Alice and Bob's system after M rounds of the protocol is given by

$$\rho_{AB}^M = \rho_{AB}^{\otimes M}, \tag{6.34}$$

i.e., they share the same state in every round of the protocol. Recall that in order to prove the security of the protocol we need to find a bound on the information that Eve has on the generated key, which is given by the min-entropy $H_{\min}^\epsilon(K_A^n|E)$ (cf. (5.83)). In the IID scenario Alice's raw key is given by $K_A^n = K_1 \ldots K_n$, where the

K_i are IID random variables. Similar, Eve's information is given by $E = E_1 \ldots E_n$, where the E_i are IID quantum side information. E_1 holds the side information about K_1, E_2 about K_2, and so on.

If we simply want to find a bound on the von Neumann entropy $H(K_A^n|K_E)$ (instead of the smooth min-entropy), we can use the chain rule from Corollary 3.12 and write

$$H(K_1 \ldots K_n|E_1 \ldots E_n) = \sum_i H(K_i|E_1 \ldots E_n K_1 \ldots K_{i-1}) \qquad (6.35)$$

$$= \sum_i H(K_i|E_i) \qquad (6.36)$$

$$= nH(K_1|E_1), \qquad (6.37)$$

where in the last line we have used that the IID assumption imposes that all $H(K_i|E_i)$ are the same. In this way, we have reduced the analysis of the entire protocol to the analysis of a single round.

A similar strategy is available for bounding the smooth min-entropy. In [43], Tomamichel et al. have proved a *Quantum Asymptotic Equipartition Property* , which states that

$$H_{\min}^\epsilon(K_A^n|E) = nH(K_1|E_1) - c_\epsilon\sqrt{n}, \qquad (6.38)$$

where c_ϵ is a correction term that is independent of n. This reduces the problem of bounding the smooth min-entropy of the whole protocol to a single-round analysis.

However, these simplifications only work because we have made the rather strong assumption that the devices are IID. In particular, this requires that the devices are memory-less, otherwise the individual rounds would not be independent. This is a strong assumption on the internal workings of the devices, which are exactly the kind of assumptions we try to avoid in DIQKD. Therefore, it is inevitable to prove the security of a protocol against the most general attacks, namely coherent attacks.

6.2.2 Security Against Coherent Attacks

Proving the security of a DIQKD protocol against coherent attacks, which are the most general attacks an eavesdropper can perform, is more difficult than proving security against collective attacks. The reason for this is that when performing coherent attacks Eve can act differently in each round and so can the devices. In the device-dependent scenario there are several techniques that reduce the security proof for coherent attacks to a security proof for collective attacks, for example, de Finetti-type theorems or the postselection technique. In the device-independent scenario these techniques can no longer be applied since they require that the Hilbert space dimension of the parties' systems in known, which is obviously not fulfilled in the device-independent setting.

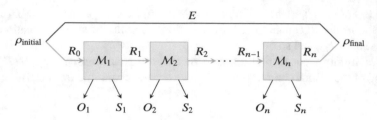

Fig. 6.6 DIQKD protocol with sequential interaction. Every round of the protocol is represented by a quantum channel M_i that takes a quantum state R_{i-1} as input and outputs classical data O_i and S_i as well as a quantum state R_i which is the input to the next round M_{i+1} of the protocol. The system is entangled with its purifying system E

As explained above, it is inevitable to lose the IID assumption and instead allow for sequential interactions of the device. This means that there can be interactions between the individual rounds of the protocol, hence the random variable K_A^n and the quantum side information E can no longer be written in the product form used above. Even worse, the requirements for the QAEP are now no longer fulfilled. Therefore, we need an extended QAEP that also allows for sequential interactions.

Fortunately, a series of works [8, 27, 33, 46] has resulted in the formulation of the *Entropy Accumulation Theorem (EAT)* [4, 17, 18], which fills this gap in the DIQKD case. Similar to the QAEP, the EAT allows to reduce the analysis of the whole protocol to that of a single round. A detailed analysis of device-independent security using the EAT can be found in [5].

The basic idea of the EAT is the following: Consider a protocol with n rounds as depicted in Fig. 6.6. Here, every round i of the protocol is modelled as a channel M_i that gets a quantum state R_{i-1} (which is entangled with the purifying system E) as input and outputs two kinds of data: Some classical data O_i (which represents the outputs A_i and B_i Alice and Bob get) and S_i (Alice and Bob's respective inputs X_i and Y_i for their devices, plus possibly additional side information) as well as a quantum state R_i, which is the input for the subsequent channel (i.e., the subsequent round of the protocol). From the classical data O_i and S_i Alice and Bob can determine whether they won or lost the CHSH game. From the outputs O_i they want to generate a secret key.

The EAT then provides a bound on the entropy by relating it to the worst case that can happen in each individual round of the protocol, roughly speaking. To understand this concept, consider a single round i of the protocol isolated from the remaining rounds. The global state of the system, denoted σ, consists of the input state R_{i-1} to the channel M_i and the purifying system R' at this point. As explained above, the output of the channel M_i consists of some classical data O_i, S_i and a quantum state R_i. The conditional von Neumann entropy is hence evaluated for the states $(M_i \otimes \mathbb{I}_R)(\sigma)$, abbreviated by $M_i(\sigma)$:

$$H(O_i|S_i R')_{M_i(\sigma)}. \tag{6.39}$$

Since we do not work in the IID setting, the local state σ is not accessible to us, so we actually have to take the minimum over all possible states σ that are compatible with the observed input-output statistics. The entropy accumulation theorem then states that the min-entropy is bounded by[7]

$$H_{\min}^{\epsilon}(O^n|S^n E)_{\rho} \geq \sum_{i=1}^{n} \min_{\sigma} H(O_i|S_i R')_{\sigma} - O(\sqrt{n}). \qquad (6.40)$$

The bound has a similar structure to the asymptotic equipartition theorem stated in (6.38): the first term is linear in the number n of rounds and we have an additional term proportional to $\sqrt{n}$, which vanishes in the limit of large n (since the expression is divided by n to compute the key rate).

6.2.3 Modifications of the Protocol

Recently, some modifications of the standard protocol have been explored on the theoretical front of DIQKD which aim at improving the secret key rate: While in the original protocol described above only one pair of measurements $\{A_0, B_1\}$ is used for key generation, it is also possible to use two randomly chosen key generation bases. This approach was studied in [38], where the authors find that this results in higher secret key rates, especially in the regime of small Bell violation (which is the regime that is accessible with the state-of-the-art experimental equipment, see Sect. 7.4).

Another approach to improve the secret key rate which was studied in [48] is to use a modified version of the CHSH inequality where different weights are given to Alice's measurements, resulting in a family of Bell inequalities given by

$$S_{\mathrm{mod}} = \alpha\langle A_1 B_1\rangle + \alpha\langle A_1 B_2\rangle + \langle A_2 B_1\rangle - \langle A_2 B_2\rangle, \qquad (6.41)$$

where $\alpha \in \mathbb{R}$ is a free parameter (note that $\alpha = 1$ corresponds to the standard CHSH inequality). A similar approach was studied in [39]. With this modification the robustness of the key generation protocol can be improved. For instance, for the case where the noise is modelled by the depolarizing channel it is possible to push the tolerable error rate (i.e., the error rate up to which we have a positive secret key rate) from 7.15% to 7.34%.

These modified DIQKD protocols naturally require more sophisticated proof methods. In [41], the authors present a framework to analyse the finite-key statistics which is applicable to general DIQKD protocols, including the modifications mentioned above.

---

[7]The actual theorem is a bit more complicated and involves some additional technical details, but the version presented here is sufficient to understand the idea.

6.3   Loopholes

As explained above, the security of device-independent security entirely relies on the violation of a Bell inequality. When performing experimental tests of such a Bell violation one has to be careful as they can be subject to one or more of several loopholes, which would imply that they in principle admit a local description.

What is required to perform a loophole-free Bell test? In general, we need to assure that the following two properties are fulfilled:

1. No information about the input of one party is allowed to be known to the other party before the output is produced.
2. The detection efficiencies have to be sufficiently high.

Loopholes arise when one or both of these requirements are not fulfilled in the experimental design or setup, which affects the validity of the results.

6.3.1   The Locality Loophole

If the first requirement is not fulfilled, we are in the situation that the premises needed for the validity of a Bell inequality are not given. This means that it is possible to find a classical model that accounts for the apparent non-locality of the observed correlations. For example, some "hidden signal" can inform Bob's experiment about Alice's choice of measurement basis and alter the state of Bob's system, hence creating quantum-like correlations. This is called the *locality loophole* [10].

A way to account for this problem is by arranging the experiment in a way such that no sub-luminal information about Alice's choice of measurement can reach Bob's system until after his measurement is complete (and vice versa), as depicted in Fig. 6.7. For this purpose, Alice and Bob have to be separated far enough and the measurements have to be performed fast enough such that no light-speed communication can affect the respective measurements.

To give Alice and Bob enough time to choose their respective measurement basis and carry out the measurement, it is necessary to at least separate them by several tens of meters. Hence, one needs to be able to send entangled states over such distances without doing much damage to them. This is why in these experiments, scientists often use entangled pairs of photons to close the locality loophole (see, for example, the experiments reported in [6,36,37,42,47]). However, these experiments suffer from insufficiencies in handling and detecting single photons, which gives rise to another loophole: the detection loophole.

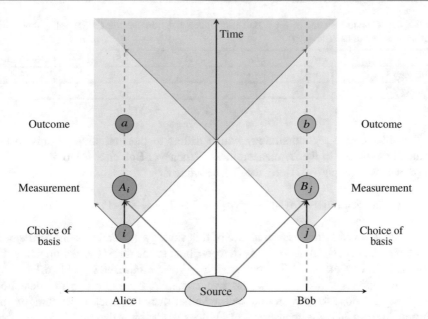

Fig. 6.7 Space-time diagram of an experimental setup that closes the locality loophole. The diagonal lines in the space-time diagram represent light-speed trajectories. The source sends entangled states to Alice and Bob (thick grey arrows). The forward light cone of the measurement choice i of Alice (j of Bob) is denoted in red (blue). The diagram shows that Bob cannot receive any information about Alice's choice of measurement until after his measurement is complete (and vice versa)

6.3.2 The Detection Loophole

The detection loophole [29] stems from the fact that, in practice, the devices that detect and measure entangled photons are imperfect. It is possible that, even though the source has sent one half of a pair of entangled photons to each party, one of the detectors simply does not respond. This can be exploited by an adversary to manipulate the input-output statistics that Alice and Bob observe.

How does this loophole affect the security of a protocol? Consider the following strategy: The four possible measurements A_0, A_1 and B_0, B_1 with possible outcome set $\{-1, +1\}$ that are used for the Bell test agree to always output the result $+1$. The value of the CHSH polynomial is then

$$\langle a_0 b_0 \rangle + \langle a_0 b_1 \rangle + \langle a_1 b_0 \rangle - \langle a_1 b_1 \rangle = 1 + 1 + 1 - 1 = 2, \qquad (6.42)$$

which reveals that this is a classical strategy and no quantum entanglement is involved. We will call this *Strategy 1*. Note that the measurement pairs that contribute to the CHSH violation are $A_0 B_0$, $A_0 B_1$, and $A_1 B_0$, while the pair $A_1 B_1$ lowers the CHSH value.

Table 6.1 Classical strategies for a Bell test. Taken individually, the CHSH value shows the classical nature of each strategy

	A_0	B_0	A_1	B_1	CHSH value
Strategy 1	+1	+1	+1	+1	2
Strategy 2	+1	+1	+1	−1	2

Consider now a second strategy which differs to the first strategy only in the value of the output of the B_1 measurement: Whenever Bob chooses to measure B_1, the device will output -1. Here, the CHSH value is

$$\langle a_0 b_0 \rangle + \langle a_0 b_1 \rangle + \langle a_1 b_0 \rangle - \langle a_1 b_1 \rangle = 1 - 1 + 1 - (-1) = 2, \tag{6.43}$$

which again reveals the classical nature of this strategy. We call this one *Strategy 2*. Here, the pairs $A_0 B_0$, $A_1 B_0$, and $A_1 B_1$ contribute to the CHSH violation while the pair $A_0 B_1$ lowers the value. The two strategies are summarized in Table 6.1.

What happens if we combine the two strategies by exploiting the detection loophole? Suppose the detectors decide whether or not to click depending on the chosen strategy and the measurement choices made by Alice and Bob. In other words, the detectors only click when the measurement choices are "good" for the CHSH violation. More precisely, they pursue the following strategy:

1. When choosing Strategy 1, Alice's detector only clicks if she chooses to measure A_0. Otherwise, there is no click.
2. When choosing Strategy 2, Alice's detector only clicks if she chooses to measure A_1. Otherwise, there is no click.

With the statistics they obtain, Alice and Bob compute the CHSH value to be

$$\langle a_0 b_0 \rangle + \langle a_0 b_1 \rangle + \langle a_1 b_0 \rangle - \langle a_1 b_1 \rangle = 1 + 1 + 1 - (-1) = 4, \tag{6.44}$$

i.e., there is a Bell violation even though the underlying strategies are classical. Note that this is only possible because Alice's detector decides whether or not to click depending Alice's measurement choice. Of course, a CHSH value of 4 is suspicious since the upper bound is given by $S \le 2\sqrt{2}$, but still you get the idea of this attack. It is possible to combine more than two classical strategies to choose between, hence constructing a CHSH value of $2 < S \le 2\sqrt{2}$, such that Alice and Bob believe that they share some entanglement even though the underlying probability distributions are classical.

They can circumvent this problem by adding a third possible outcome, the "no click"-event, which means that the set of possible outcomes is $\{-1, +1, \text{"no click"}\}$. The resulting probability distribution is then in the classical region. However, this gives rise to a new problem with regard to practical implementations: For the resulting probability distribution to be in the quantum region, the amount of "no click"-events (i.e., photon losses) has to be below 5–10%, which is difficult to achieve since in order to close the locality loophole, the devices need to be at a certain distance.

Exercise 6.13 Show that when using the combined strategy described above, taking into account the "no click"-outcome as a possible measurement outcome yields a CHSH value in the classical regime.

To overcome the detection loophole, the obvious strategy is to use a different source of entanglement than photons. For instance, entangled qubits are slower and heavier and can be measured with success probability close to 1. This strategy was pursued by various groups (see, for example, [35] and [24]) using trapped ions, superconducting qubits, and other systems. However, with these kinds of systems it is impossible to overcome long distances due to decoherence of the states. As a consequence, the individual parts of these experiments have to be close to each other, which in turn leaves the locality loophole open. Hence, for a long time experimental Bell tests were able to either close the locality loophole or the detection loophole, but not both at the same time.

A more practical way to overcome the detection loophole and the locality loophole both at the same time is the following: Shortly before the photon arrives at Bob's laboratory, it undergoes a Quantum Non-Demolition (QND) measurement where it has to decide whether it is going to click or not, as depicted in Fig. 6.8. The QND basically asks the photon, "Photon, are you there?", i.e., it measures the presence of the photon without disturbing it. Why is this better than the previous scenario? Before, the photon decided whether or not to click dependent on Bob's input. With the QND, the photon now has to say whether or not it is going to click without knowing Bob's input. Alice and Bob then only perform a Bell test in case the photon is arriving and all the losses that happen before are irrelevant for the Bell test statistics.

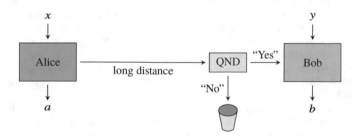

Fig. 6.8 Bell test with Quantum Non-Demolition (QND) measurement. Right before the photon arrives at Bob's laboratory, it undergoes a QND which checks whether the photon is there or has been lost. Only if there is a photon, Alice and Bob proceed with performing the Bell test

References

1. Acín, A., Massar, S., Pironio, S.: Efficient quantum key distribution secure against no-signalling eavesdroppers. New J. Phys. **8**(8), 126–126 (2006). https://doi.org/10.1088/1367-2630/8/8/126
2. Acín, A., Brunner, N., Gisin, N., Massar, S., Pironio, S., Scarani, V.: Device-independent security of quantum cryptography against collective attacks. Phys. Rev. Lett. **98**(23), 230501 (2007). https://doi.org/10.1103/physrevlett.98.230501
3. Arnon-Friedman, R.: Device-Independent Quantum Information Processing: A Simplified Analysis. Springer International Publishing, New York (2020). https://doi.org/10.1007/978-3-030-60231-4
4. Arnon-Friedman, R., Dupuis, F., Fawzi, O., Renner, R., Vidick, T.: Practical device-independent quantum cryptography via entropy accumulation. Nat. Commun. **9**(1) (2018). https://doi.org/10.1038/s41467-017-02307-4
5. Arnon-Friedman, R., Renner, R., Vidick, T.: Simple and tight device-independent security proofs. SIAM J. Comput. **48**(1), 181–225 (2019). https://doi.org/10.1137/18m1174726
6. Aspect, A., Dalibard, J., Roger, G.: Experimental test of Bell's inequalities using time-varying analyzers. Phys. Rev. Lett. **49**(25), 1804–1807 (1982). https://doi.org/10.1103/physrevlett.49.1804
7. Barrett, J., Hardy, L., Kent, A.: No signaling and quantum key distribution. Phys. Rev. Lett. **95**(1) (2005). https://doi.org/10.1103/physrevlett.95.010503
8. Barrett, J., Colbeck, R., Kent, A.: Unconditionally secure device-independent quantum key distribution with only two devices. Phys. Rev. A **86**(6) (2012). https://doi.org/10.1103/physreva.86.062326
9. Bell, J.S.: On the Einstein Podolsky Rosen paradox. Phys. Phys. Fiz. **1**(3), 195–200 (1964). https://doi.org/10.1103/physicsphysiquefizika.1.195
10. Bell, J.S.: Speakable and Unspeakable in Quantum Mechanics. Cambridge University Press, Cambridge (1987). https://doi.org/10.1017/cbo9780511815676
11. Cirel'son, B.S.: Quantum generalizations of Bell's inequality. Lett. Math. Phys. **4**(2), 93–100 (1980). https://doi.org/10.1007/bf00417500
12. Clauser, J.F., Horne, M.A., Shimony, A., Holt, R.A.: Proposed experiment to test local hidden-variable theories. Phys. Rev. Lett. **23**(15), 880–884 (1969). https://doi.org/10.1103/physrevlett.23.880
13. Cleve, R., Hoyer, P., Toner, B., Watrous, J.: Consequences and limits of nonlocal strategies. In: Proceedings. 19th IEEE Annual Conference on Computational Complexity. IEEE (2004). https://doi.org/10.1109/ccc.2004.1313847
14. Colbeck, R.: Quantum and relativistic protocols for secure multi-party computation. Ph.D. thesis, University of Cambridge (2006)
15. Colbeck, R., Renner, R.: Free randomness can be amplified. Nat. Phys. **8**(6), 450–453 (2012). https://doi.org/10.1038/nphys2300
16. Devetak, I., Winter, A.: Distillation of secret key and entanglement from quantum states. Proc. Roy. Soc. A Math. Phys. Eng. Sci. **461**(2053), 207–235 (2005). https://doi.org/10.1098/rspa.2004.1372
17. Dupuis, F., Fawzi, O.: Entropy accumulation with improved second-order term. IEEE Trans. Inform. Theory **65**(11), 7596–7612 (2019). https://doi.org/10.1109/tit.2019.2929564
18. Dupuis, F., Fawzi, O., Renner, R.: Entropy accumulation. Commun. Math. Phys. **379**(3), 867–913 (2020). https://doi.org/10.1007/s00220-020-03839-5
19. Ekert, A.K.: Quantum cryptography based on Bell's theorem. Phys. Rev. Lett. **67**(6), 661–663 (1991). https://doi.org/10.1103/physrevlett.67.661
20. Freedman, S.J., Clauser, J.F.: Experimental test of local hidden-variable theories. Phys. Rev. Lett. **28**(14), 938–941 (1972). https://doi.org/10.1103/physrevlett.28.938
21. Gallego, R., Masanes, L., Torre, G.D.L., Dhara, C., Aolita, L., Acín, A.: Full randomness from arbitrarily deterministic events. Nat. Commun. **4**(1) (2013). https://doi.org/10.1038/ncomms3654

22. Giustina, M., Versteegh, M.A., Wengerowsky, S., Handsteiner, J., Hochrainer, A., Phelan, K., Steinlechner, F., Kofler, J., Larsson, J.Å., Abellán, C., Amaya, W., Pruneri, V., Mitchell, M.W., Beyer, J., Gerrits, T., Lita, A.E., Shalm, L.K., Nam, S.W., Scheidl, T., Ursin, R., Wittmann, B., Zeilinger, A.: Significant-loophole-free test of Bell's theorem with entangled photons. Phys. Rev. Lett. **115**(25) (2015). https://doi.org/10.1103/physrevlett.115.250401

23. Hensen, B., Bernien, H., Dréau, A.E., Reiserer, A., Kalb, N., Blok, M.S., Ruitenberg, J., Vermeulen, R.F.L., Schouten, R.N., Abellán, C., Amaya, W., Pruneri, V., Mitchell, M.W., Markham, M., Twitchen, D.J., Elkouss, D., Wehner, S., Taminiau, T.H., Hanson, R.: Loophole-free Bell inequality violation using electron spins separated by 1.3 kilometres. Nature **526**(7575), 682–686 (2015). https://doi.org/10.1038/nature15759

24. Matsukevich, D.N., Maunz, P., Moehring, D.L., Olmschenk, S., Monroe, C.: Bell inequality violation with two remote atomic qubits. Phys. Rev. Lett. **100**(15) (2008). https://doi.org/10.1103/physrevlett.100.150404

25. Mayers, D., Yao, A.: Quantum cryptography with imperfect apparatus. In: Proceedings 39th Annual Symposium on Foundations of Computer Science. IEEE (1998). https://doi.org/10.1109/sfcs.1998.743501

26. Miller, C.A., Shi, Y.: Robust protocols for securely expanding randomness and distributing keys using untrusted quantum devices. In: Proceedings of the Forty-Sixth Annual ACM Symposium on Theory of Computing. ACM (2014). https://doi.org/10.1145/2591796.2591843

27. Miller, C.A., Shi, Y.: Robust protocols for securely expanding randomness and distributing keys using untrusted quantum devices. J. ACM **63**(4), 1–63 (2016). https://doi.org/10.1145/2885493

28. Nielsen, M.A., Chuang, I.L.: Quantum Computation and Quantum Information. Cambridge University Press, Cambridge (2000)

29. Pearle, P.M.: Hidden-variable example based upon data rejection. Phys. Rev. D **2**(8), 1418–1425 (1970). https://doi.org/10.1103/physrevd.2.1418

30. Pironio, S., Acín, A., Brunner, N., Gisin, N., Massar, S., Scarani, V.: Device-independent quantum key distribution secure against collective attacks. New J. Phys. **11**(4), 045021 (2009). https://doi.org/10.1088/1367-2630/11/4/045021

31. Pironio, S., Acín, A., Massar, S., de la Giroday, A.B., Matsukevich, D.N., Maunz, P., Olmschenk, S., Hayes, D., Luo, L., Manning, T.A., Monroe, C.: Random numbers certified by Bell's theorem. Nature **464**(7291), 1021–1024 (2010). https://doi.org/10.1038/nature09008

32. Reichardt, B.W., Unger, F., Vazirani, U.: Classical command of quantum systems. Nature **496**(7446), 456–460 (2013). https://doi.org/10.1038/nature12035

33. Reichardt, B.W., Unger, F., Vazirani, U.: A classical leash for a quantum system. In: Proceedings of the 4th Conference on Innovations in Theoretical Computer Science - ITCS '13. ACM Press (2013). https://doi.org/10.1145/2422436.2422473

34. Rosenfeld, W., Burchardt, D., Garthoff, R., Redeker, K., Ortegel, N., Rau, M., Weinfurter, H.: Event-ready Bell test using entangled atoms simultaneously closing detection and locality loopholes. Phys. Rev. Lett. **119**(1), 010402 (2017). https://doi.org/10.1103/physrevlett.119.010402

35. Rowe, M.A., Kielpinski, D., Meyer, V., Sackett, C.A., Itano, W.M., Monroe, C., Wineland, D.J.: Experimental violation of a Bell's inequality with efficient detection. Nature **409**(6822), 791–794 (2001). https://doi.org/10.1038/35057215

36. Salart, D., Baas, A., van Houwelingen, J.A.W., Gisin, N., Zbinden, H.: Spacelike separation in a Bell test assuming gravitationally induced collapses. Phys. Rev. Lett. **100**(22), 220404 (2008). https://doi.org/10.1103/physrevlett.100.220404

37. Scheidl, T., Ursin, R., Kofler, J., Ramelow, S., Ma, X.S., Herbst, T., Ratschbacher, L., Fedrizzi, A., Langford, N.K., Jennewein, T., Zeilinger, A.: Violation of local realism with freedom of choice. Proc. Natl. Acad. Sci. USA **107**(46), 19708–19713 (2010). https://doi.org/10.1073/pnas.1002780107

38. Schwonnek, R., Goh, K.T., Primaatmaja, I.W., Tan, E.Y.Z., Wolf, R., Scarani, V., Lim, C.C.W.: Device-Independent Quantum Key Distribution with random key basis. Nat Commun. **12** (1), 1–8 (2021). https://doi.org/10.1038/s41467-021-23147-3

39. Sekatski, P., Bancal, J.D., Valcarce, X., Tan, E.Y.Z., Renner, R., Sangouard, N.: Device-Independent Quantum Key Distribution from Generalized CHSH Inequalities (2020). Preprint: https://arxiv.org/abs/2009.01784
40. Shalm, L.K., Meyer-Scott, E., Christensen, B.G., Bierhorst, P., Wayne, M.A., Stevens, M.J., Gerrits, T., Glancy, S., Hamel, D.R., Allman, M.S., Coakley, K.J., Dyer, S.D., Hodge, C., Lita, A.E., Verma, V.B., Lambrocco, C., Tortorici, E., Migdall, A.L., Zhang, Y., Kumor, D.R., Farr, W.H., Marsili, F., Shaw, M.D., Stern, J.A., Abellán, C., Amaya, W., Pruneri, V., Jennewein, T., Mitchell, M.W., Kwiat, P.G., Bienfang, J.C., Mirin, R.P., Knill, E., Nam, S.W.: Strong loophole-free test of local realism. Phys. Rev. Lett. **115**(25), 250402 (2015). https://doi.org/10.1103/physrevlett.115.250402
41. Tan, E.Y.Z., Sekatski, P., Bancal, J.D., Schwonnek, R., Renner, R., Sangouard, N., Lim, C.C.W.: Improved DIQKD Protocols with Finite-Size Analysis (2020). Preprint: https://arxiv.org/abs/2012.08714
42. Tittel, W., Brendel, J., Zbinden, H., Gisin, N.: Violation of Bell inequalities by photons more than 10 km apart. Phys. Rev. Lett. **81**(17), 3563–3566 (1998). https://doi.org/10.1103/physrevlett.81.3563
43. Tomamichel, M., Colbeck, R., Renner, R.: A fully quantum asymptotic equipartition property. IEEE Trans. Inform. Theory **55**(12), 5840–5847 (2009). https://doi.org/10.1109/tit.2009.2032797
44. Vazirani, U., Vidick, T.: Certifiable quantum dice: or, true random number generation secure against quantum adversaries. In: Proceedings of the 44th Symposium on Theory of Computing - STOC '12. ACM Press (2012). https://doi.org/10.1145/2213977.2213984
45. Vazirani, U., Vidick, T.: Fully device-independent quantum key distribution. Phys. Rev. Lett. **113**(14), 140501 (2014). https://doi.org/10.1103/physrevlett.113.140501
46. Vazirani, U., Vidick, T.: Robust device independent quantum key distribution. In: Proceedings of the 5th Conference on Innovations in Theoretical Computer Science - ITCS '14. ACM Press (2014). https://doi.org/10.1145/2554797.2554802
47. Weihs, G., Jennewein, T., Simon, C., Weinfurter, H., Zeilinger, A.: Violation of Bell's inequality under strict Einstein locality conditions. Phys. Rev. Lett. **81**(23), 5039–5043 (1998). https://doi.org/10.1103/physrevlett.81.5039
48. Woodhead, E., Acín, A., Pironio, S.: Device-Independent Quantum Key Distribution Based on Asymmetric CHSH Inequalities (2020). Preprint: https://arxiv.org/abs/2007.16146

Recent Developments in Practical QKD

7

Abstract

Turning an abstract protocol into a practical device offers a variety of challenges. The main problem is that the devices used in the implementation such as optical fibres and single-photon detectors are never perfect. For instance, one has to take into account that photons get lost when sent though an optical fibre, especially over long distances. In this chapter, we discuss the challenges that arise when practically implementing QKD schemes. We furthermore present developments in QKD that aim at overcoming these practical challenges, such as measurement device-independent QKD and continuous-variable QKD, and give a brief overview of the state-of-the art experiments for several types of protocols.

7.1 Practical Challenges in QKD

We begin by discussing the main practical challenges that arise when implementing a QKD protocol. From a historical point of view, considering practical issues has often led to ground-breaking discoveries: For example, in order to find a counter-attack to the photon-number-splitting attack the decoy state protocol was invented, which nowadays is a standard protocol in practical QKD. With regard to the performance of a QKD protocol and its suitability for a certain application, several aspects have to be considered:

1. **Distance.** An important criterion of whether the protocol is useful for a certain application is at what distance a secure key can be generated. Depending on the protocol and on the kind of implementation there are different limiting factors. For instance, in optical experiments a crucial limitation is the photon loss in optical fibres and terrestrial free-space.
2. **Key rate.** Since key generation is the ultimate goal of the QKD implementation, the secure key rate that can be generated is an important measure of performance

© The Author(s), under exclusive license to Springer Nature Switzerland AG 2021   183
R. Wolf, *Quantum Key Distribution*, Lecture Notes in Physics 988,
https://doi.org/10.1007/978-3-030-73991-1_7

that has to be taken into account. A practical implementation of a QKD protocol over long distances is useless if it does not yield a positive secret key rate. The long-term goal here is to close the gap between classically achievable rates and QKD communication rates. Classically, rates of 100 Gbit/s are currently deployed while QKD schemes can achieve rates in the Mbit/s regime. While this is already sufficient for video transmission, for example, the long-term goal is to be able to encrypt high volumes of classical network traffic. The obtained key rates crucially depend on the detectors that are employed. For instance, in schemes that use single-photon detectors high detection efficiencies and a short dead time are essential for achieving high key rates.

3. **Security proof.** Apart from being able to generate keys at a high rate over great distances, it is further important that there exists a composable security proof for the underlying protocol that includes security against general attacks and also takes into account finite-size effects. Further aspects of practical security include, for example, side channel attacks against the detectors, which represent critical weaknesses of QKD implementations that can be exploited by an eavesdropper as discussed in the next section. On the classical side of the protocol, it is crucial that efficient methods exist for post-processing the raw key bits since we usually have to deal with large blocks of data.

With regard to commercial applications it is furthermore important that the cost of the devices is not too high alongside with a high performance. Important questions are, for instance, whether QKD systems can coexist with intense data traffic in the same optical fibre or whether the system needs to be cooled, which are both factors that have an influence on the cost of the system. More details on these aspects can be found in [13].

One aspect of practical QKD that is not covered in this book, but worth mentioning, is the scenario where secure communication between more than two parties is required. It is of course possible to trivially achieve security in this case by performing two-party QKD protocols between pairs of parties and use the resulting keys to distribute the conference key. An alternative to this procedure is the so-called *multipartite QKD* or *quantum conference key agreement (CKA)*. Here, one exploits the correlations that arise in multipartite entangled states to create a truly multipartite QKD protocol that directly establishes a secret key between all involved parties. These kinds of protocols are the first step towards quantum networks and generalizations of the BB84 protocol and the Six-state protocol have already been proposed in [27] and [14], respectively. We do not go into more detail about CKA here, but it is covered in detail in [26] and a review of the topic can be found in [54].

7.2   Measurement Device-Independent QKD

One of the most vulnerable parts of a QKD implementation is the detectors (see, for example, [44]). Even in a device-independent setting, where we do not make any assumptions on the internal workings of the devices, the setting is still vulnerable to the so-called *detector side channel attacks*, as described in more detail in the next section. To circumvent this problem, Measurement Device-Independent QKD (MDO QKD) rules out the possibility of such attacks by removing all detectors from the trusted part of the setting (i.e., Alice and Bob's labs) and moves them to an untrusted relay. Alice and Bob then both only prepare and send quantum states instead of receiving them.

7.2.1 Detector Side Channel Attacks

There is a variety of side channel attacks that exploit the imperfectness of single-photon detectors. To get an idea of why these detectors are especially vulnerable we give two examples of such attacks, the *time-shift attack* and the *detector blinding attack*.

Time-Shift Attack
Usually in QKD it is assumed that the detection efficiencies for the two bits "0" and "1" are equal. However, this assumption is not necessarily justified as shown in [49, 50, 59], for example. Perfectly matching detector efficiencies can only be guaranteed if they are constant in time, with is impossible in practice, for example, because of their intrinsic dead time.

How can Eve exploit this detection efficiency mismatch? Consider the scenario depicted in Fig. 7.1, where the efficiencies of the detectors are shifted slightly in time. Since Eve can manipulate the photons sent by Alice, she can shift the arrival time of a photon randomly to either point A or point B with probability p_A and $p_B = 1 - p_A$, respectively. She chooses p_A carefully such that Bob still gets an equal number of "0" and "1" outcomes. Since Bob's measurement result is shifted

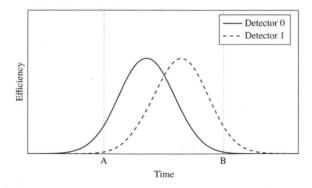

Fig. 7.1 Detector efficiency mismatch. The plot shows a typical time dependence of the detection efficiencies of a practical fibre-based QKD system. If the efficiencies for the two detectors that correspond to the bits "0" and "1", respectively, are not the same, an eavesdropper can perform the time-shift attack to gain information on the key bits

towards one of the outcomes depending on the time shift (A or B), Eve can get information without alarming Alice and Bob.

This kind of attack can even be implemented with current technology, as shown in [85]. Hence, one does not even have to give the eavesdropper unlimited power as it is done in standard security proofs. A weak eavesdropper that is limited to today's technology is already capable of performing this attack.

Detector Blinding Attack

In experiments, often avalanche photodiodes (APDs) are used to detect single-photon pulses. The detector blinding attack exploits the fact that if bright light is shined onto the detectors they become insensitive to single-photon pulses and only detect strong light pulses. In this way, Eve can effectively control which detector produces a click by sending additional bright pulses to Bob.

To understand how this attack works, we first have to understand the internal workings of an avalanche photodiode, which is depicted in Fig. 7.2. An APD has two modes of operation: The *Geiger mode*, which is used for single-photon detection, and the *linear mode*, in which the detector is insensitive to single-photon pulses (see Fig. 7.3). When the APD is in Geiger mode, the voltage inside the APD is above the breakdown voltage V_{br}, hence it is ready to detect a single photon. An incoming single-photon pulse will release an avalanche of photons, which yields a large current I_{APD} in the APD. This eventually causes the detector to click when the current I_{APD} is above the threshold I_{th}. After the click the voltage inside the APD is lowered beyond V_{br} to quench the avalanche. The APD is then set back into Geiger mode, awaiting the next single-photon pulse.

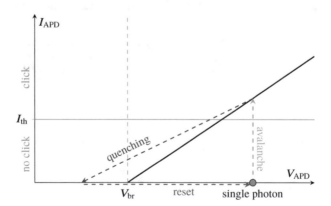

Fig. 7.2 Working principle of an avalanche photodiode (APD) in Geiger mode. In this mode the APD is able to detect single-photon pulses: An incoming photon releases an avalanche inside the photodiode which results in a large current I_{APD}. When the current in the APD is above the threshold, i.e., $I_{APD} > I_{th}$, the detector clicks. Afterwards, the voltage inside the APD is lowered beyond the breakdown voltage V_{br} in order to quench the avalanche. Then the APD is set back into Geiger mode

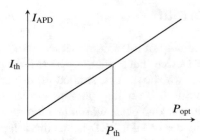

Fig. 7.3 APD in linear mode. If the voltage V_{APD} inside the APD is below the breakdown voltage V_{br}, the APD operates in linear mode, where it is unable to detect single-photon pulses: The current I_{APD} in the APD is linear to the incident optical power P_{opt}, making the current threshold I_{th} an optical power threshold P_{th}

When the voltage in the APD is below V_{br}, the photodiode is in linear mode. This means that the current I_{APD} is proportional to the incident optical power P_{opt}. In this mode the threshold I_{th} becomes an optical power threshold P_{th}. The detector is then unable to detect single-photon pulses. It only clicks above a certain intensity threshold.

By shining bright light onto the detectors, Eve can make them operate in linear mode instead of Geiger mode. Then she can perform an intercept-and-resend attack:

1. Eve detects the states sent by Alice in a random basis.
2. She resends her detection results to Bob, but instead of using single-photon pulses she sends bright pulses with an intensity that is just above the optical power threshold P_{th}.
3. Bob only gets a click if his measurement choice coincides with Eve's choice.

After the quantum transmission phase of the protocol, Eve and Bob hold identical bit values and basis choices. Since classical post-processing is done via open classical communication, Eve can listen to it and apply the same operations as Bob to her key bits, thus obtaining an identical final key.

This attack works for all implementations that use APD-based detectors, which is the vast majority of commercial and research QKD systems, and was demonstrated in several works, see, for example, [17, 46, 48].

The attacks explained above are just two of several possible detector side channel attacks. For instance, one can also exploit the intrinsic dead time of single-photon detectors (for example, the time it takes to quench and reset an APD), which was studied in [76]. Another possibility that falls under the name *Trojan horse attack* [22, 73] is based on the idea of shining bright light into Alice's or Bob's setup to extract information on the internal settings. Measuring the reflection of the pulse might allow Eve to collect information on the basis choices Alice and Bob have done, for example.

7.2.2   Practical MDI QKD

To overcome the vulnerability of a QKD implementation caused by detector side channels, Measurement Device-Independent QKD (MDI QKD) aims at replacing detectors by transmitters, which was first studied in [43] and [6]. More precisely, in an MDI QKD scheme Bob is sending photons instead of detecting them. The detection procedure is then taking place within an intermediate relay that connects Alice and Bob. The basic idea of MDI QKD is that this relay is untrusted, and since all detectors involved are placed in the intermediate relay, none of them have to be trusted. Hence, detector side channel attacks do not alter the security of the protocol, in contrast to schemes where the detection process takes place at Bob's lab.

Even though Eve can have full control of the untrusted relay, it is still possible for Alice and Bob to establish a secret key if the outcome of the measurement, which is publicly announced, is informative for Alice and Bob, but does not reveal any information about the key to Eve. This is, for example, possible if a Bell measurement is done at the untrusted relay. Before we discuss in detail how the key generation works, we study a possible practical implementation of this scheme using an optical setup (following [43]) as depicted in Fig. 7.4. Each round of the protocol works as follows:

1. Alice and Bob each prepare a phase-randomized weak coherent pulse.
2. A polarization modulator prepares the pulse in one of the four polarization states known from the BB84 protocol by randomly selecting the polarization, which encodes a random bit in the polarization of the pulse: in the rectilinear basis, where we have horizontal (H) and vertical (V) polarization, we use the encoding $0 \rightarrow H$, $1 \rightarrow V$, while in the diagonal basis with $+45°$ (D) and $-45°$ (A) polarization the encoding is $0 \rightarrow D$, $1 \rightarrow A$.
3. An intensity modulator modifies the amplitude of the pulse, thereby generating either signal or decoy states in order to rule out the possibility of Eve performing the photon-number-splitting attack without being detected.
4. The two pulses are sent to an untrusted relay where they first interfere at a 50:50 beam splitter.
5. At each output port of the beam splitter is a polarization beam splitter which projects the pulse into either horizontal or vertical polarization states.
6. Four single-photon detectors (two on each side) are employed to detect the photons. The detection results are publicly announced.

The four detectors can discriminate two of the four Bell states, which can be seen by carrying out a virtual qubit ansatz: suppose Alice prepares an entangled state between the state she is sending, which is an either horizontally or vertically polarized photon, and a virtual qubit she is holding. We denote the polarized photon by using the Fock state formalism, for example, a single photon that is horizontally polarized is denoted $|1\rangle_{A_H}$. The virtual qubit contains information on the polarization of the single photon and is denoted either $|H\rangle$ or $|V\rangle$, which

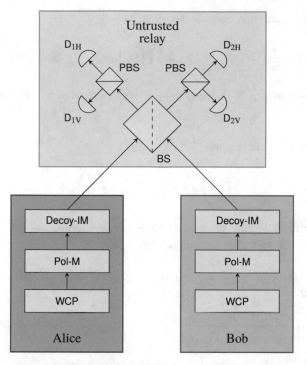

Fig. 7.4 Scheme of an MDI QKD experiment. Alice and Bob both have a source of phase-randomized weak coherent pulses (WCP). These pulses are sent through a polarization modulator (Pol-M) which encodes a random bit into the polarization of the pulse. An intensity modulator (Decoy-IM) is used to implement the decoy state method. The two pulses are then sent to an untrusted relay, where they first interfere at a 50 : 50 beam splitter (BS) followed by two polarization beam splitters (PBS), one at each output port, which project the incoming pulses into either horizontal (H) or vertical (V) polarization states. In the end, four single-photon detectors (D_{1V}, D_{1H}, D_{2V}, D_{2H}) are used to detect the pulses, effectively doing a Bell measurement

corresponds to the computational basis as explained above. The same considerations are made for Bob's system; hence, the resulting states of the preparation process are

$$|\psi_A\rangle = \frac{1}{\sqrt{2}} \left(|H\rangle_A |1\rangle_{A_H} + |V\rangle_A |1\rangle_{A_V} \right) \tag{7.1}$$

$$|\psi_B\rangle = \frac{1}{\sqrt{2}} \left(|H\rangle_B |1\rangle_{B_H} + |V\rangle_B |1\rangle_{B_V} \right). \tag{7.2}$$

Measuring the virtual qubit in either the rectilinear or the diagonal basis is then equivalent to preparing one of the four BB84 states. However, since the measurements commute with the detection process at the untrusted relay, they can be delayed until after the photon detection has occurred. Note that we can write the Fock states as $|1\rangle_{A_H} = a_H^\dagger |0\rangle$ using the respective creation operator. The global state after the preparation phase is the given by

$$|\psi_A\rangle \otimes |\psi_B\rangle = \frac{1}{2}\left(|H\rangle_A a_H^\dagger + |V\rangle_A a_V^\dagger\right)\left(|H\rangle_B b_H^\dagger + |V\rangle_B b_V^\dagger\right)|0\rangle \tag{7.3}$$

$$= \frac{1}{2}\left(|HH\rangle_{AB} a_H^\dagger b_H^\dagger + |VV\rangle_{AB} a_V^\dagger b_V^\dagger + |HV\rangle_{AB} a_H^\dagger b_V^\dagger + |VH\rangle_{AB} a_V^\dagger b_H^\dagger\right)|0\rangle. \tag{7.4}$$

Using the beam splitter relation [19]

$$\begin{pmatrix} a^\dagger \\ b^\dagger \end{pmatrix} \mapsto \frac{1}{\sqrt{2}}\begin{pmatrix} 1 & 1 \\ 1 & -1 \end{pmatrix}\begin{pmatrix} c^\dagger \\ d^\dagger \end{pmatrix}, \tag{7.5}$$

we can rewrite (7.4) and, after combining suitable terms, we get the following expression for the state after passing the beam splitter:

$$|\psi_{BS}\rangle = \frac{1}{2}\left(|HH\rangle_{AB}\frac{|2\rangle_{1H} - |2\rangle_{2H}}{2} + |VV\rangle_{AB}\frac{|2\rangle_{2V} - |2\rangle_{2V}}{2}\right. \tag{7.6}$$

$$\left. + |\Psi^+\rangle_{AB}\frac{|1\rangle_{1H}|1\rangle_{1V} - |1\rangle_{2H}|1\rangle_{2V}}{\sqrt{2}} - |\Psi^-\rangle_{AB}\frac{|1\rangle_{1H}|1\rangle_{2V} - |1\rangle_{1V}|1\rangle_{2H}}{\sqrt{2}}\right),$$

where $|\Psi^+\rangle_{AB} = \frac{1}{\sqrt{2}}(|HV\rangle_{AB} + |VH\rangle_{AB})$ and $|\Psi^-\rangle_{AB} = \frac{1}{\sqrt{2}}(|HV\rangle_{AB} - |VH\rangle_{AB})$, and $|1\rangle_{1H}$ denotes a single horizontally polarized photon at the left output port (see Fig. 7.4).

From (7.6), we can directly see that a successful Bell measurement can occur at most $1/2$ of the time, which reduces the key rate. More precisely, it occurs for four combinations of detector clicks: the click of D_{1H} and D_{2V} or D_{1V} and D_{2H} indicates a projection into $|\Psi^-\rangle$, while the click of D_{1H} and D_{1V} or D_{2H} and D_{2V} indicates a projection into $|\Psi^+\rangle$.

The remaining possible outcomes cannot be used for key generation: if both photons are detected at the same detector, the polarizations of both of them are revealed, which in turn means that the bits Alice and Bob hold are also revealed. Note that a third case, where the photons arrive at either D_{1H} and D_{2H} or D_{1V} and D_{2V} does not occur in the formula (7.6). This is because of the *Hong-Ou-Mandel effect* (HOM) [33], which says that if two identical photons enter the 50 : 50 beam splitter together they will always leave at the same output. The occurrence of the HOM effect is important to the efficiency of the MDI scheme since it reduces the number of possible detection patterns at the relay, hence making a successful Bell measurement more likely.

These results are publicly announced, and Alice and Bob only keep those data that corresponds to rounds where a successful Bell measurement has taken place. Furthermore, as in the BB84 protocol, Alice and Bob post-select those events where they have used the same basis in the preparation step, using an authenticated public classical channel. To successfully generate a key, Alice and Bob need perfectly correlated bits. If they have used the rectilinear basis for encoding, a successful Bell measurement only happens if their bits are different (see (7.6)). They can agree that Bob flips his bit in these cases. If they use the diagonal basis, the measurement result $|\Psi^+\rangle$ indicates that their bits are the same, while the result $|\Psi^-\rangle$ indicates that their bits are different, hence Bob flips his bit (see Exercise 7.2). In summary, the measurement result only provides the information of the parity of their bits, but not the bit values. Hence, this information is useful for Alice and Bob (since they know the values of their individual bits), but not for Eve.

Exercise 7.1 Use the beam splitter relation (7.5) to derive the expression of the state after passing the beam splitter (7.6) from Eq. (7.4).

Exercise 7.2 Suppose Alice and Bob have used the diagonal basis for encoding their states instead of the rectilinear basis.

1. Suppose Alice has chosen the diagonal basis to encode a 0-bit, which means that she sends the state $|\psi_A\rangle = |1\rangle_{A_D} = \frac{1}{\sqrt{2}}\left(|1\rangle_{A_H} + |1\rangle_{A_V}\right)$, while Bob encodes a 1-bit into the diagonal basis:[1] $|\psi_B\rangle = |1\rangle_{B_A} = \frac{1}{\sqrt{2}}\left(|1\rangle_{A_H} - |1\rangle_{A_V}\right)$. Show that if the Bell measurement is successful it always yields $|\Psi^-\rangle$ (and never $|\Psi^+\rangle$). *Hint: Use the beam splitter relation (7.5).*
2. Show that if Alice and Bob encode the same bit in the diagonal basis, the photons will always end up on the same side of the beam splitter (this is the HOM effect), which means the only possible Bell measurement yields $|\Psi^+\rangle$.

These calculations show how Alice and Bob can generate perfectly correlated bits when using the diagonal basis for encoding: if the result is $|\Psi^-\rangle$, they know that their bits are *different*. They can then agree that Bob flips his bit. If the result is $|\Psi^-\rangle$, they know that their bits are the same.

Security of MDI QKD

The security proof of the MDI QKD protocol [43] described above relies on the security proof of the BB84 protocol with weak coherent pulses by Gottesmann, Lo, Lütkenhaus, and Preskill [25] and combines ideas from the security proof of a time-reversed EPR-based QKD protocol [34] and the decoy state method.

In the decoy state setting Alice and Bob use the single-photon events to estimate the QBER and the gain (i.e., the probability that the relay outputs a successful result). We use the notation $e_{\text{rect}}^{n,m}$, $e_{\text{diag}}^{n,m}$ to denote the QBER in the rectilinear and diagonal basis, respectively, where n and m denote the number of photons sent by Alice and Bob and given that the relay produces a successful output. In the same way we denote the gain by $Q_{\text{rect}}^{n,m}$ and $Q_{\text{diag}}^{n,m}$.

---

[1] We have to be a bit careful with the notation here: $|1\rangle_{B_A}$ means a single photon in Bob's system that is polarized $-45°$ (A). The A here has nothing to do with Alice.

Using the virtual qubit setting, the protocol described above is equivalent to an entanglement-based protocol since we can think of it as Alice and Bob getting handed a pair of entangled qubits, which ideally is either in the state $|\Psi^+\rangle$ or in the state $|\Psi^-\rangle$. Alice and Bob then measure their qubits either in the rectilinear or diagonal basis. Suppose they use the results from measuring in the rectilinear basis for key generation, while the events where they measured in the diagonal basis are used to estimate the knowledge Eve has (i.e., for parameter estimation). For an ideal setup and in the asymptotic limit of an infinitely long key, the secret key rate is then simply given by the probability that a successful output is produced by the relay in case Alice and Bob have both prepared a single photon in the rectilinear basis:

$$r_{\text{MDI}}^{\text{ideal}} = Q_{\text{rect}}^{1,1}. \tag{7.7}$$

If we take into account that in a practical implementation there are imperfections such as basis misalignment and detector dark counts, the asymptotic key rate is given by

$$r_{\text{MDI}} = Q_{\text{rect}}^{1,1}\left(1 - h_2\!\left(e_{\text{diag}}^{1,1}\right)\right) - Q_{\text{rect}} f(E_{\text{rect}}) h_2(E_{\text{rect}}), \tag{7.8}$$

where $Q_{\text{rect}} = \sum_{n,m} Q_{\text{rect}}^{n,m}$ and $E_{\text{rect}} = \sum_{n,m} Q_{\text{rect}}^{n,m} E_{\text{rect}}^{n,m}/Q_{\text{rect}}$ are the total gain and QBER, respectively, in the rectilinear basis, and $f(E_{\text{rect}}) > 0$ is an inefficiency function for the error correction process.[2] As usual, h_2 denotes the binary Shannon entropy. The quantities $Q_{\text{rect}}^{1,1}$ and $e_{\text{diag}}^{1,1}$ can be bounded using the decoy state method. The finite-key analysis for this protocol was carried out in [9]. Here, the authors showed that even though finite-size effects reduce both the achievable distance and the secret key rate, it is possible to perform secure MDI QKD over up to 150 km with a finite set of data, say of 10^{12} to 10^{14} signals.

Achievable Distances
Apart from being secure under all detector side channel attacks, MDI QKD has the advantage that its long-distance performance is better than that of conventional QKD. To see this, let us do a simple analysis of signal and noise in the respective schemes. Consider a standard QKD protocol where Alice prepares and sends states and Bob measures them. Let us denote the probability that a single photon sent by Alice arrives at the detector by η_{chan} (also known as the transmittance of the channel). Further, denote by η_{det} the probability that Bob's detector clicks given that a photon has arrived at the detector and denote by p_{dark} the probability that a dark count occurs (i.e., the detector clicks even though no photon is present). We

---

[2]Note that the effect of the inefficiency function is usually very small. In [43], the authors used a value of $f(E_{\text{rect}}) = 1.16$ for simulation purposes but pointed out that this is a rather conservative choice. With good error correction codes one can achieve better results.

then define a *signal S* as

$$S = \eta_{\text{chan}}\eta_{\text{det}}, \tag{7.9}$$

while the *noise N* is simply the probability p_{dark} that a dark count occurs. We could now say that generating a secure key is not possible if the noise-to-signal ratio is

$$\frac{N}{S} \gtrsim 10\%, \tag{7.10}$$

for example.[3] Suppose now that we have a detection efficiency of $\eta_{\text{det}} = 50\%$ and a probability of dark counts of $p_{\text{dark}} = 10^{-5}$. From the condition (7.10) it follows that the channel needs to have a transmittance of at least $\eta_{\text{chan}} = 2 \times 10^{-4}$. Furthermore, we can use the following relation between the transmittance and the length of the fibre to determine the maximum length of the fibre:

$$\eta_{\text{chan}} = 10^{-l_{\text{dB}}L/10}. \tag{7.11}$$

Here, l_{dB} denotes the transmission loss in the fibre. For standard fibres that are used in optical experiments, a typical value is $l_{\text{dB}} = 0.2\,\text{dB/km}$ (this means that about half the photons are lost after $1.5\,\text{km}^4$) [21]. From (7.11) we then get that the maximum length of the fibre is $L_{\text{max}} = 185\,\text{km}$.

For MDI QKD, signal and noise are defined slightly different. The first difference is that the fibre now consists of two parts, the one that connects Alice with the relay and the one that connects Bob with the relay. To achieve the same overall distance, the length of each of the fibres is half of the overall distance, which means the each of the fibres has a transmittance of $\sqrt{\eta_{\text{chan}}}$. The second difference is that we now have two detectors instead of one, which yields the following expression for the signal S:

$$S_{\text{MDI}} = \sqrt{\eta_{\text{chan}}}\sqrt{\eta_{\text{chan}}}\eta_{\text{det}}\eta_{\text{det}} = \eta_{\text{chan}}\eta_{\text{det}}^2. \tag{7.12}$$

Also, the expression for the noise is more complicated for MDI QKD: since we have two detectors, we have to consider the probability of dark counts twice, and also the probability that a dark count only occurs at one of the detectors while the other one clicks (which can happen for Alice's detector as well as for Bob's). Hence, the expression for the noise is given by

$$N_{\text{MDI}} = p_{\text{dark}}^2 + 2p_{\text{dark}}\sqrt{\eta_{\text{chan}}}\eta_{\text{det}}. \tag{7.13}$$

---

[3]The actual number depends on the devices that are used in the experimental implementation, but for our purposes it is not important as long as we choose the same number in the conventional QKD setting and in the MDI QKD setting.

[4]The losses in dB and the losses in % are related by the formula $l_{\text{dB}} = -10\log_{10}(1 - (l_\%/100))$.

Using the same conditions as above, we get that the minimum value for the transmittance is $\eta_{\text{chan}} = 1.68 \times 10^{-7}$ and the maximum distance is $L_{\text{max}} = 339\,\text{km}$. Even though this is just a rough analysis, it indicates that with MDI QKD it is possible to achieve much greater distances than with conventional QKD schemes. This is because with the intermediate relay, the noise in the channels is handled better than if one long channel connects the two parties.

Exercise 7.3 Derive (7.11). Use the relation between the loss in dB and the loss in % given in Footnote 4 taking into account that the total loss in the fibre is given by $l_{\text{dB}} \cdot L$, where L is the length of the fibre. *Hint: Use that the transmittance η_{chan} is defined as the probability that a photon is successfully transmitted.*

7.2.3 Twin-Field QKD

While MDI QKD overcomes the vulnerability to side channel attacks it is subject to a limitation from which all point-to-point QKD schemes suffer: there is a fundamental limitation for the secure key rate due to the losses that occur when transmitting quantum states via a lossy channel [56, 71]. In particular, the secret key rate of a QKD protocol over a quantum channel with transmittance η is upper bounded by the so-called PLOB bound (named after its discoverers Pirandola, Laurenza, Ottaviani, and Banchi) [56]:

$$r \leq -\log(1 - \eta). \tag{7.14}$$

In the regime of high distances, that is, high losses $\eta \ll 1$, the logarithm can be expanded and we find that

$$r \approx 1.44\eta, \tag{7.15}$$

i.e., the maximum achievable secret key rate scales linear with the transmittance of the channel.

A lot of research has been done to explore the possibility of using quantum repeaters to overcome this limitation, see [67]. However, this approach has the disadvantage that it requires quantum memory and quantum error correction and is therefore beyond the capabilities of current technology.

Another approach to overcome the rate-distance limit (7.15) was presented in 2018 by Lucamarini et al. [45] and is called *Twin-Field QKD (TF QKD)*. The authors propose a scheme that is based on similar principles as MDI QKD: Alice and Bob send pulses to an untrusted relay at which the measurement takes place. The measurement results reveal the parity of the encoded bits, but not their values. Different to MDI QKD, TF QKD only uses single-photon detection, which results in a higher key rate since successful events now correspond to the situation where one photon arrives, sent either by Alice or Bob, in contrast to the requirement of having a successful Bell measurement at the relay (where both photons are needed).

Fig. 7.5 Phase slices for a
TF QKD protocol.
Discretization of the phase
space for $M = 16$ to identify
the twin fields in the public
discussion

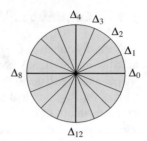

The protocol presented in [45] works as follows: in the encoding step (which is exactly the same for Alice and Bob, so we only describe Alice's part here), Alice generates phase-randomized weak coherent pulses by picking a random phase value $\varphi_a \in [0, 2\pi)$. The phase-interval is split into M slices with $\Delta_k = 2\pi k/M$, where $k = 0, \ldots, M - 1$, see Fig. 7.5. The phase φ_a falls into one of these intervals, which we denote $\Delta_{k(a)}$. Alice then encodes a secret bit and a secret basis into another phase ψ_a which is added to the phase of the pulse.

At the relay, the two incoming pulses (one from Alice, one from Bob) are combined at a 50 : 50 beam splitter with single-photon detectors at the output ports. After the detection outcome is announced, Alice and Bob reveal their respective phase slices $\Delta_{k(a)}$ and $\Delta_{k(b)}$ and the encoded bases. They only keep the data that corresponds to rounds with matching values. The optical fields whose phases are from the same slice are "twins", hence the name twin-field QKD. The detection outcome together with the revealed information tells Bob whether or not he has to flip his bit to coincide with Alice's bit. An eavesdropper, on the other hand, only has the information of the parity of Alice and Bob's bits but cannot learn their absolute value.

Note that while a high number M of slices gives a small QBER, the probability of having matching slides scales as $1/M$, hence there is an optimal value for M with regard to the secret key rate. In [45], the authors determined this value to be $M_{\text{opt}} = 16$ by modelling the experimental setup and optimizing the key rate.

As explained in the beginning, the goal of TF QKD is to overcome the rate-distance limit of point-to-point QKD given in (7.15) by achieving a quadratic improvement, i.e., a scaling of $O(\sqrt{\eta})$ instead of $O(\eta)$. In the original paper [45], the authors were able to prove such an improvement for a restricted class of attacks by Eve. Later works by Tamaki et al. [72] and Ma et al. [47] showed the security of modified versions of the original protocol against general attacks while also achieving a $O(\sqrt{\eta})$ scaling. A simplified version of the original protocol, presented by Curty et al. [10], uses pre-selection of the phases instead of postselection. This modification simplifies the security analysis and is easier to implement experimentally while also achieving a square root improvement over the point-to-point rate-distance limitation.

7.3   Continuous-Variable QKD

Historically, much of the theory (and also of the experiments) on quantum communication has originally been developed for Discrete Variables (DV), which is what we have focussed on so far. The reason behind this is that by using discrete variables a lot of the intuition and ideas from classical information science could be adapted to the quantum case. However, in the 2000s it became clear that also continuous variables of the electromagnetic field, such as the amplitude and phase of light, are suitable candidates for quantum communication tasks, (see, e.g., [61] and [28]). In fact, using protocols based on continuous variables can be advantageous with regard to practical applications since the states and measurements that are involved in the theoretical description, such as coherent states and homodyne detection, can be directly used in an experimental implementation. In contrast to this, single-photon sources used in discrete-variable protocols, for example, have to be approximated by weak coherent pulses. In this section we give a basic introduction over the most important concepts of Continuous-Variable (CV) QKD an discuss the advantages compared to discrete-variable QKD as well as the current challenges in CV QKD.

7.3.1   Basics of Gaussian Quantum Information

The main idea of CV QKD is to encode the information into the degrees of freedom of the quantized electromagnetic field. To understand how protocols built on this idea work we first have to introduce some basic concepts of quantum optics. Most of what we discuss in this section can be found in the detailed review on this topic by Weedbrook et al. [75].

In the discrete-variable setting the fundamental building block of our protocols was the qubit, often realized as the polarization of a single photon. Hence, the first question that naturally comes up is what is the analogue of a qubit in the continuous setting? In CV QKD, we do not use the notion of a qubit but rather work with modes. The analogue quantity would then be a "qumode", which is a quantized mode of a bosonic system. A prominent example for this is the degrees of freedom of the electromagnetic field. In Table 7.1, important quantities of QKD protocols are compared for DV and CV QKD.

Description of Quantum Systems
The first and fundamental difference between DV and CV quantum systems is that the Hilbert space is infinite. The CV quantum system can be represented by N quantized modes of the electromagnetic field, i.e., N bosonic modes. This corresponds to N quantized Harmonic oscillators. Each of these harmonic oscillators can be described by quadrature field operators $\hat{q}$ and $\hat{p}$, which act like the position and momentum of a harmonic oscillator. For the i-th oscillator, they are

Table 7.1 Discrete-variable vs. continuous-variable QKD. This table shows the analogue quantities in CV QKD to what we have used in the description of DV QKD protocols

	DV QKD	CV QKD
Light	Photons (or WCS)	Wave
Quantities of interest	Number and coherence	Amplitude and phase or quadratures $\hat{q}$, $\hat{p}$
Description	Density matrix ρ	Wigner function $W(\hat{q}, \hat{p})$
Measurements	Counting (e.g., with an APD)	Demodulating: homodyne or heterodyne detection
"Simple" states	Fock states	Gaussian states

defined as

$$\hat{q}_i = \hat{a}_i^\dagger + \hat{a}_i, \tag{7.16}$$

$$\hat{p}_i = i(\hat{a}_i^\dagger - \hat{a}_i), \tag{7.17}$$

where $\hat{a}_i$ and $\hat{a}_i^\dagger$ are the creation and annihilation operators, respectively, of the bosonic field. The information we want to send is encoded in the quadrature field operators $\hat{q}$ and $\hat{p}$, which are continuous variables: let us write all operators $\hat{q}_i$, $\hat{p}_i$ in one vector $\hat{\mathbf{x}} = (\hat{q}_1, \hat{p}_1, \ldots, \hat{q}_N, \hat{p}_N)^\mathsf{T}$. The eigenvalues $\mathbf{x}$ of this vector of operators are given by

$$\hat{\mathbf{x}}^\mathsf{T} |\mathbf{x}\rangle = \mathbf{x}^\mathsf{T} |\mathbf{x}\rangle, \tag{7.18}$$

where $|\mathbf{x}\rangle = (|x_1\rangle, \ldots |x_{2N}\rangle)^\mathsf{T}$. The vector $\mathbf{x}$ of eigenvalues of the quadrature field operators contains of $2N$ real numbers, hence they form *continuous variables* that describe the entire bosonic system.

For a single mode, the quadrature field operators fulfil the commutation relation

$$[\hat{q}, \hat{p}] = 2i N_0, \tag{7.19}$$

where N_0 is a normalization factor that is also known as *shot noise* and which is sometimes set to one. This factor also appears in the uncertainty relation of the two operators,

$$\Delta \hat{q} \Delta \hat{p} \geq N_0 \tag{7.20}$$

and as such it describes the minimum variance that is reachable symmetrically by the two quadrature field operators. This naturally leads to the optical phase space picture of states as depicted in Fig. 7.6 for a coherent state. The marginal probability distributions can be obtained by measuring one of the two quadratures. Because of the uncertainty relation (7.20), we cannot measure both quadratures with arbitrary

Fig. 7.6 Optical phase space of a coherent state. A coherent state corresponds to a blurred dot in the optical phase space since, because of Heisenberg's uncertainty relation, not both field quadratures $\hat{q}$ and $\hat{p}$ can be measured simultaneously with arbitrary precision

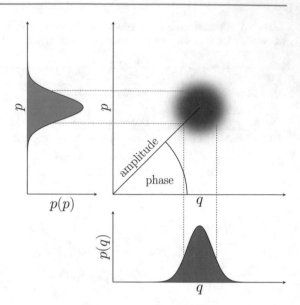

precision, hence a coherent state cannot be represented by a sharp point in phase space.

An equivalent description of a CV state ρ is the Wigner function $W(\hat{\mathbf{x}})$, which is defined on the $2N$-dimensional phase space. It is a *quasi-probability distribution*, which means that it has all the properties of a probability distribution except that it can take negative vales for non-classical states.[5] The Wigner function is characterized by the statistical moments of the quantum state: the first moment is called the displacement vector or mean value

$$\bar{\mathbf{x}} = \langle \hat{\mathbf{x}} \rangle = \mathrm{Tr}\left(\hat{\mathbf{x}}\rho\right). \tag{7.21}$$

The second moment is called the covariance matrix $\mathbf{V}$ whose elements are given by

$$V_{ij} = \frac{1}{2}\langle\{\Delta\hat{x}_i, \Delta\hat{x}_j\}\rangle, \tag{7.22}$$

where $\Delta\hat{x}_i = \hat{x}_i - \langle\hat{x}_i\rangle$ and $\{\cdot, \cdot\}$ is the anti-commutator. In particular, this means that the diagonal elements of the covariance matrix provide the variance of the quadrature operators:

$$V_{ii} = V(\hat{x}_i) = \langle\Delta\hat{x}_i^2\rangle = \langle\hat{x}_i^2\rangle - \langle\hat{x}_i\rangle^2. \tag{7.23}$$

The covariance matrix is real, symmetric, and positive definite.

---

[5]Note that the converse is not true in general: a state can be non-classical yet have a non-negative Wigner function.

For a particular class of states the mean value and the covariance matrix are sufficient to provide a complete description of the state of the system, i.e., $\rho = \rho(\bar{\mathbf{x}}, \mathbf{V})$. These states are the so-called *Gaussian states* which are defined as bosonic states whose Wigner function is Gaussian. Why are we interested in Gaussian states with regard to the implementation of cryptographic protocols? The reason is twofold: first, Gaussian states are very robust with interaction of the environment and Gaussian channels (i.e., CPTP maps that map Gaussian states to Gaussian states) describe a lossy optical fibre, which is the standard case in practical QKD based on photonic devices. The second reason why we are interested in Gaussian states is that it is known how to generate them experimentally using lasers, for instance.

The Wigner function for a Gaussian state is given by

$$W(\mathbf{x}) = \frac{1}{(2\pi)^N (\det \mathbf{V})^{1/2}} e^{-\frac{1}{2}(\mathbf{x}-\bar{\mathbf{x}})^\mathsf{T} \mathbf{V}^{-1}(\mathbf{x}-\bar{\mathbf{x}})}, \tag{7.24}$$

where $\mathbf{x}$ is the vector of eigenvalues of the quadrature operators $\hat{\mathbf{x}}$. From the above formula it is obvious that we do not need any higher-order moments than the mean and the variance to determine the Wigner function.

Examples of Gaussian states are the eigenstates of the harmonic oscillator, i.e., Fock states, which are depicted in Fig. 7.7 in terms of the Wigner function. Further interesting examples are coherent states:

$$|\alpha\rangle = e^{-\frac{|\alpha|^2}{2}} \sum_n \frac{\alpha^n}{\sqrt{n!}} |n\rangle. \tag{7.25}$$

They can be generated from the vacuum by applying the displacement operator $D(\alpha)$, which is a Gaussian unitary operation, via $|\alpha\rangle = D(\alpha)|0\rangle$, where

$$D(\alpha) = e^{\alpha \hat{a}^\dagger - \alpha^* \hat{a}} \tag{7.26}$$

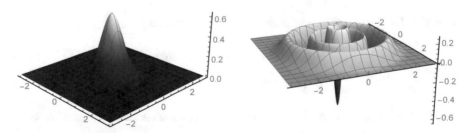

Fig. 7.7 Wigner distribution for eigenstates of the harmonic oscillator. The plots show the Wigner distribution for the ground state $|0\rangle$ (left hand side) and the Fock state $|5\rangle$ (right hand side). On the horizontal axes are the values of the quadratures q and p. In the plot on the right hand side it becomes clear that the Wigner distribution can become negative and is therefore a quasi-probability distribution

is the displacement operator on a single mode and α is a complex number. It transforms the mode operator $\hat{a}$ as

$$D^{\dagger}(\alpha)\hat{a}D(\alpha) = \hat{a} + \alpha. \tag{7.27}$$

Note that this operation leaves the variance of the Wigner function invariant. Coherent states are non-orthogonal, i.e.,

$$|\langle\beta|\alpha\rangle|^2 = e^{-|\beta-\alpha|^2}. \tag{7.28}$$

This is important for the security of CV QKD since non-orthogonal quantum states cannot be distinguished with certainty.

Another prominent example is *squeezed states*, which are generated from the vacuum by a squeezing operation $S(s)$ (which is also Gaussian unitary operation) and a displacement operation:

$$|\alpha, s\rangle = D(\alpha)S(s)|0\rangle. \tag{7.29}$$

Here, the variance of one quadrature operator is squeezed while the variance of the other quadrature operator is expanded. Both coherent and squeezed states can be generated experimentally and are therefore useful for implementing CV QKD protocols. For simplicity, we only focus on coherent state in the following.

Before we turn to the description of measurements, a remark on the use of coherent states is necessary. We have already seen coherent states in the description of discrete-variable protocols, for example, weak coherent states in the decoy state method. It is important to understand that the use of coherent states in DV QKD is fundamentally different from their use in CV QKD. In DV QKD, they are used to approximate single photons and the information is encoded in the polarization of the single photon. In CV QKD, the information is instead encoded in the quadrature field operators, i.e., we use the degrees of freedom of the field itself and not of a single photon.

Exercise 7.4 Use (7.27) to show the following identities:

$$D^{\dagger}(\alpha)\hat{q}D(\alpha) = \hat{q} + \sqrt{2}\,\mathrm{Re}(\alpha) \tag{7.30}$$

$$D^{\dagger}(\alpha)\hat{p}D(\alpha) = \hat{p} + \sqrt{2}\,\mathrm{Im}(\alpha). \tag{7.31}$$

Exercise 7.5 Calculate the mean value and the covariance matrix for the vacuum state $|0\rangle$ of a single-mode system.

Exercise 7.6 Calculate the mean value and the covariance matrix for a coherent state $|\alpha\rangle = D(\alpha)|0\rangle$. *Hint: Use that the displacement operator preserves variances.*

Measurements

The aim of measuring a Gaussian state is to get information about the quadrature field operators. When measuring a Gaussian state the outcomes are distributed according to a Gaussian probability distribution and as such they are elements of the real numbers (in contrast to the discrete outcomes that we have in DV QKD, for instance, whether a detector clicks or not).

Typical examples of suitable measurements are *homodyne* and *heterodyne* detection. When doing homodyne detection only one of the quadratures ($\hat{q}$ or $\hat{p}$) is measured. The measurement operators of a homodyne detection are projectors over the quadrature basis $|q\rangle\langle q|$ or $|p\rangle\langle p|$ (depending on which quadrature is measured). The probability of getting an outcome q (or p, respectively) is given by the marginal integral of the Wigner function over the other quadrature:

$$P(q) = \int W(q, p)dp, \quad P(p) = \int W(q, p)dq. \tag{7.32}$$

Experimentally, homodyne detection is implemented by comparing the incoming signal with a standard oscillation that corresponds to the signal if it carried null information, as depicted in Fig. 7.8. The two signals interfere at a 50:50 beam splitter and two detectors measure the intensity of the outgoing modes. Subtracting the outcomes of the two detectors gives a signal which is proportional to $\hat{q}$. To measure the $\hat{p}$ quadrature, a $\pi/2$ phase shift is applied to the local oscillator [7]. This kind of measurement is near-optimal with regard to distinguishing low-intensity coherent states.

With heterodyne detection, on the other hand, we can measure both quadratures at once at the cost of getting a noise penalty due to the uncertainty relation (7.20). While in the case of homodyne detection Bob needs to have access to a random number generator in order to randomly choose which quadrature to measure, this is not required when using heterodyne detection. As a result, the experimental setup

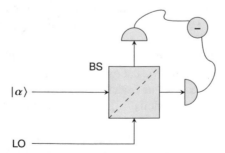

Fig. 7.8 Homodyne detection. The incoming signal $|\alpha\rangle$ interferes with a reference signal provided by the local oscillator LO at a 50:50 beam splitter BS. Two photon detectors measure the outcomes at the two output ports of the beam splitter and the homodyne current is then given by difference of the detector outputs

is simpler using heterodyne detection and it furthermore produces higher key rates (see [74]).

7.3.2   CV QKD Protocols

We have now discussed all the basics required to understand how a CV QKD protocol works. In general, there are two ingredients to every protocol: an encoding of Gaussian states (i.e., a modulation) and a decoding using Gaussian measurements (usually homodyne or heterodyne detection). Many different CV QKD protocols have been proposed in the literature which differ in one or more of the following aspects:

1. **State preparation:** One can use protocols with single-mode or two-mode coherent or squeezed states.
2. **Modulation:** The encoding of information into Gaussian states can be achieved via Gaussian modulation (where the states are distributed according to a Gaussian distribution) or non-Gaussian modulation.
3. **Detection:** The measurement can be done with either homodyne or heterodyne detection.
4. **Post-processing:** One can choose between direct or reverse information reconciliation, which makes a big difference in CV QKD, and also between one-way or two-way communication.

These possibilities yield a variety of protocols, where some protocols are easier to implement while others have better security proofs.

As it is usually the case in QKD, protocols can be described in two different ways: in the prepare-and-measure (PM) version and in the entanglement-based (EB) version, which are equivalent for Gaussian protocols [30]. In the PM case, Alice prepares Gaussian states and sends them to Bob, who measures them with either homodyne or heterodyne detection. In the EB case, Alice prepares bipartite entangled Gaussian states, measures the first half and sends the second half to Bob, who then measures it. Given that Alice's lab and preparation are trusted, both versions of the protocol yield the same security.

To get familiar with the workings of CV QKD protocols we present one of the simplest protocols, which is a one-way protocol using coherent states with Gaussian modulation and homodyne detection known as the GG02 protocol [28] (or its version with heterodyne detection [74]). The steps of the protocol are as follows:

1. **State distribution and measurement:** Alice prepares N coherent states $|\alpha_1\rangle$, $\ldots$, $|\alpha_N\rangle$, where the α_i are IID variables that are distributed according to a Gaussian distribution with zero mean and variance V_A. The states are then sent to Bob. Depending on the protocol, Bob measures either one of the quadratures ($\hat{q}$ or $\hat{p}$) at random for each of the states (homodyne detection) and tells Alice his choice or he measures both quadratures (heterodyne detection). Afterwards, Bob

holds either a list of N or $2N$ real-valued numbers according to the measurement outcomes. Alice keeps only the relevant quadrature values according to Bob's measurement choices. Denote the resulting lists $x = (x_1, \ldots, x_n)$ for Alice and $y = (y_1, \ldots, y_n)$ for Bob, where $n = N$ or $n = 2N$ according to the measurement type.

2. **Error correction:** The next step in the protocol is to correct errors in Alice and Bob's respective bit strings. Here, it turns out that reverse reconciliation gives better results for these types of protocols where we have high-loss channels [29]. This means that Bob's string corresponds to the raw key and Alice tries to guess its value, in contrast to direct reconciliation that we have used in the protocols previously discussed. The error correction itself is then a classical procedure, where Alice and Bob use an error correction code they have agreed on beforehand.

3. **Parameter estimation:** Before, we have usually placed the parameter estimation step before error correction. However, it has turned out that the other order can be more efficient. In CV QKD, this step requires estimating the covariance matrix of the bipartite state shared by Alice and Bob to estimate the correlations they share.

4. **Privacy amplification:** This step is similar to what we have discussed in the DV setting. Alice and Bob apply a random two-universal hash function to their respective corrected strings and get a pair of secure keys S_A, S_B of length l.

A simple scheme of the protocol is depicted in Fig. 7.9. We now discuss some of the technical details of the protocol and important parameters. The state preparation that takes place in Alice's lab is characterized by the variance V_A of the Gaussian

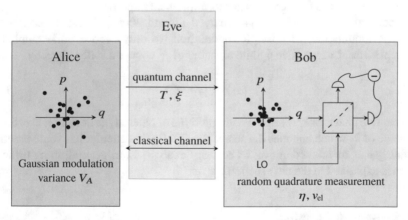

Fig. 7.9 Implementation of the GG02 protocol. Alice prepares a large number of coherent states according to a Gaussian distribution with variance V_A (Gaussian modulation). She sends the states over a quantum channel with transmission efficiency T and excess noise ξ. Due to the noise in the channel the variance of the states Bob receives is smaller than V_A. Bob then performs a homodyne detection to measure one of the quadratures at random. His measurement is described by the efficiency η and the electric noise v_{el}

modulation. The next step, sending the states to Bob over a quantum channel, is characterized by two parameters of the channel: The transmission efficiency T (i.e., the probability that a signal is successfully transmitted) and the excess noise ξ, which represents all noise due to the channel that goes beyond the shot noise. The measurement process that takes place in Bob's lab is characterized by the parameters of the detectors: the quantum efficiency η and the electric noise v_{el}. In summary, the protocol can be characterized by five parameters: V_A, T, ξ, η, and v_{el}.

These parameters allow us to study the effect on the quantum state when going through the quantum channel. Denote $X_A \in \mathcal{N}_{\mathbb{C}}(0, V_A)$ the Gaussian variable that describes the state Alice has prepared. After the state has gone through the quantum channel and has been detected in Bob's lab, the outcome is described by the Gaussian variable

$$X_B = \sqrt{\eta T}\,(X_A + X_N), \tag{7.33}$$

where X_N is a noise variable with zero mean and variance $V_N = N_0 + \eta T \xi + v_{\mathrm{el}}$. Hence, the variance of X_B is

$$V_B = \eta T\,(V_A + \xi) + N_0 + v_{\mathrm{el}}. \tag{7.34}$$

Discrete Modulation

The first attempts of CV QKD protocols around 2000 were realized using a discrete (hence, a non-Gaussian) encoding of Gaussian states, see [32, 60, 63]. However, with the discovery that Gaussian modulation of coherent states provides a good resource for continuous-variable protocol, the idea of discrete modulation had been somewhat forgotten for a while, with only a few papers studying its feasibility. Because discrete modulation is easier to implement experimentally and has higher error correction efficiencies it was revived around 2009 [37, 69]. In contrast to Gaussian modulation, for discrete modulation we only need a finite number of different states, i.e., we have a finite alphabet of N coherent states given by

$$|\alpha_k\rangle = |ae^{\frac{i2\pi k}{N}}\rangle \tag{7.35}$$

with relative phase $2\pi k/N$. Bob uses homodyne or heterodyne detection to estimate the value of k, which encodes the secret key. While this attempt has some practical advantages as mentioned above, its security analysis is, in general, more difficult, which is explained in the next section.

7.3.3 Security Analysis

Similar to security proofs of DV QKD protocols, the goal of a CV QKD security proof is to verify sufficient correlations between Alice and Bob such that we get an upper bound on Eve's information about the key. However, in contrast to the discrete setting, we now have to deal with infinite Hilbert spaces and unbounded operators.

To analyse the security of a CV QKD protocol for the case of collective attacks and in the asymptotic limit of an infinitely long key, we employ the Devetak-Winter rate:

$$r_\infty^{\text{coll}} = I(A : B) - \chi(B : E), \tag{7.36}$$

where $I(A : B)$ is the mutual information between Alice and Bob and $\chi(B : E)$ is the Holevo information between Bob's string and Eve's quantum system. This is the case if reverse reconciliation is used in the error correction step. For direct reconciliation, it has to be replaced with $\chi(A : E)$. Since in a realistic setting Alice and Bob usually cannot extract all of the information from their data, we multiply the mutual information with a factor $\beta < 1$ called the *reconciliation efficiency*. Hence, the formula for the secret key rate is given by

$$r_\infty^{\text{coll}} = \beta I(A : B) - \chi(B : E). \tag{7.37}$$

The quantity $\beta I(A : B)$ can directly be observed in an experiment.

The quantity $\chi(B : E)$, on the other hand, has to be bounded using the covariance matrix $\mathbf{V}$ of the state that is shared by Alice and Bob. This is, in general, a challenging task. In the asymptotic limit there is a useful tool to bound the Holevo information: using optimality properties of Gaussian states [78] one can show that

$$\chi(B : E)_\rho \leq \chi(B : E)_{\rho_G}, \tag{7.38}$$

where ρ_G is the Gaussian state with the same covariance matrix as ρ (see [16, 55]). This means that the optimal attack that Eve can perform is based on Gaussian operations. As a consequence, the security analysis can be restricted to this case, which makes it much easier. However, this technique is restricted to protocols with Gaussian modulation and only applicable for collective attacks and in the asymptotic limit.

For general attacks the situation is more complicated. There are different attempts of proving security in this case. For instance, Furrer et al. [15] first discretise the quadratures q_δ, p_δ and then use an entropic uncertainty relation to prove security against general attacks. The drawback of this approach is that it only works for squeezed states (which are more difficult to realize experimentally) and, furthermore, the resulting bound on the secure key rate is believed not to be tight.

Another approach is to use a Gaussian version of the de Finetti theorem. Similar to the discrete version, the states used in the protocol have to fulfil some symmetry condition in order to be able to apply the de Finetti theorem. This approach was used to show that security against collective attacks implies security against general attacks in the asymptotic case of an infinite key by Renner and Cirac [64] and in the finite-key regime by Leverrier et al. [38].

A third potential approach to prove security against general attacks is developing an entropy accumulation theorem for continuous variables. However, there is a number of challenges that need to be overcome to formulate such a theorem: for instance, it requires some test (like the CHSH game) to estimate the quantum correlations between the parties. In the CV case this should be related to the covariance matrix but it is not clear how exactly this has to look. Furthermore, this test depends on some unbounded continuous outcome (in contrast to the CHSH value) which complicates the analysis.

For protocols with discrete modulation proving unconditional security is even harder. While discrete modulation has some advantages over Gaussian modulation regarding its practical implementation, the underlying theory provides some challenges which makes even the security proof against collective attacks in the asymptotic case non-trivial. Most of the tools described above cannot be used since the requirements are no longer fulfilled: For instance, since we no longer use Gaussian modulation, (7.38) does not hold and the security analysis cannot be restricted to Gaussian operations. Furthermore, the symmetry that is required for the Gaussian de Finetti theorem is broken. Hence, the security analysis of CV protocols with discrete modulation needs new techniques. Here, some progress has been made recently: in [20] and [41], the authors have used semi-definite programs (SDP) to numerically estimate bounds on the min-entropy. In another work by Matsuura et al. [52] the CV protocol is mapped to a qubit protocol and then security proof techniques from discrete-variable QKD are employed.

A summary of the state-of-the art of security proofs for different CV QKD protocols can be found in [12].

7.3.4   Advantages and Challenges

As we have seen above there are some fundamental differences between QKD with discrete variables and with continuous variables. Regarding practical implementations, using continuous quantum systems has some advantages. For instance, CV QKD experiments employ exactly those states and measurements specified in the theoretical formulation of the protocol, such as Gaussian states and homodyne or heterodyne detection. In contrast, the BB84 protocol requires a single-photon source which, in practice, can only be approximated using weak coherent states. The states that are used in CV QKD, usually coherent or squeezed states, can be generated in the lab, with coherent states being particularly easy to prepare. Furthermore, the devices needed for a CV QKD implementation such as coherent measurements are also widely used for classical communication in the telecom industry.

On the other hand, there are some drawbacks to CV QKD. On the theoretical side there are many challenges when it comes to proving the security of a protocol such as an infinite-dimensional Hilbert space and continuous and unbounded measurement operators. Moreover, the quality of correlations between the two parties is measured in terms of the covariance matrix, which is a continuous and unbounded quantity in contrast to the CHSH violation or the QBER that we employ in DV QKD security proofs. On the experimental side, the drawback of CV QKD is that it is less robust to noise: for instance, in DV QKD it is possible to discard no click events from detectors while in CV QKD all pulses are there, but noisier. This makes it harder to estimate the channel properties precisely and leads to a need for very large block sizes for long distances.

7.4 Advances in Experimental QKD

Since the days of the first implementation of a QKD protocol (namely, the BB84 protocol), which was done in 1989 by Bennett, Brassard, and others [2, 3], much progress has been made, both on the theoretical side and with regard to the experimental components. While we have discussed recent advances regarding protocols and their security such as device-independent and measurement-device-independent QKD, we have not discussed the experimental implementations yet. In this section, we give a brief overview over the state-of-the-art of current experimental implementations of the different kinds of protocols we have presented in this book. More detailed reviews of this topic can be found in [57, 79] and [13], for example.

BB84 Protocol with Decoy States
Being the first QKD protocol the BB84 protocol is naturally well-understood and many experiments have been carried out to achieve ever greater distances and ever higher key rates. As explained earlier, an implementation with photonic devices is vulnerable to the photon-number-splitting attack. Therefore, photonic experiments usually implement the BB84 protocol using decoy states (see Sect. 5.3.2).

The experiment of Yuan et al. [82] concentrated on generating high key rates over short distances and the authors have been able to achieve a secure key rate of 10 Mbit/s.[6] While this experiment demonstrates that it is possible to achieve very high secure key rates, the distance over which the key is generated is too short for practical QKD applications.

One of the most problematical issues in long-distance QKD is the loss in optical fibres. Unlike in classical telecommunications, signals cannot noiselessly be amplified due to the no-cloning theorem which limits the maximum distance for secure QKD to a few hundred kilometres [5]. Due to the exponential decrease of the photonic signal, the detector noise will eventually become the dominant source of error, which then makes it impossible to extract a key. In [4], Boaron et al. have been able to achieve a distance of up to 421 km, but only with very small key rates. For instance, over a distance 405 km the secure key rate is only 6 bit/s. This is depicted in Fig. 7.10 together with the secure key rates for other distances (see Table 7.2), showing how it decreases with growing distance.

One possible way to overcome this limitation is to use low-Earth-orbit satellites as links. Compared to terrestrial channels, satellite-to-ground communication has a

---

[6]For comparison: the state-of-the-art classical communication channels transmit around 100 Gbit/s (see, for example, [77]).

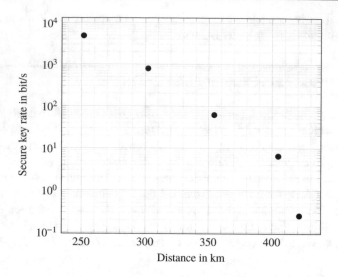

Fig. 7.10 Secret key rates over long distances. The plot shows the secret key rate for different distances listed in Table 7.2 which was achieved in an implementation of the BB84 protocol with decoy states [4]

Table 7.2 Data for secret key rates. The table lists the secret key rates for different distances achieved in an implementation of the BB84 protocol with decoy states by Boaron et al. [4], together with the observed QBER

Distance in km	QBER in %	Secret key rate in bit/s
251.7	0.5	4.9×10^3
302.1	0.4	0.79×10^3
354.5	0.7	62
404.9	1.0	6.5
421.1	2.1	0.25

greatly reduced amount of losses [62], making it a promising candidate for practical long-distance quantum communication. In 2017, two groups in China and Japan have independently demonstrated that satellite-to ground QKD is possible [39, 70]. In particular, Liao et al. [39] have been able to establish satellite-to-ground QKD over 1.200 km between the low-Earth-orbit satellite *Micius* and a ground station in Xinglong near Beijing while achieving an average key rate of 1 kbit/s.

The same satellite has later been used as a trusted relay between three different ground stations in Xinglong, Nanshan (both in China), and Graz (Austria) [40]. This works as follows: The satellite first establishes individual secret keys between itself and each of the ground stations. Upon request from the ground stations (say, Xinglong and Graz) it then performs a bitwise exclusive or-operations on the respective keys and relays the result to one of the ground stations. In this way, it is possible to establish a secure key between ground stations which are about 7.600 km apart. This demonstration included the transmission of images using a one-time pad configuration as well as a video call between Beijing and Graz.

Entanglement-Based Protocols

There has also been some progress on the implementation of entanglement-based protocols recently. A remarkable result that was reported in 2019 by Joshi et al. [35] shows a fully connected graph consisting of eight parties in a city-wide network, where entanglement-based QKD was used to establish secret keys between any pairing of the network.

Another important result with respect to the distance over which secure quantum communication is possible was reported in 2020: Yin et al. [81] achieved secure key generation over a distance of 1120 km with a key rate of 0.12 bit/s without the need for trusted relays. To generate entangled photon pairs between the two ground stations again the satellite Micius was used.

One of the greatest challenges for experimental entanglement-based protocols is achieving a high success rate in generating entangled photon states. Furthermore, the implementations are susceptible to detector side channels and imperfections of single-photon detectors.

Device-Independent QKD

The first step towards a successful implementation of device-independent protocols is to demonstrate a loophole-free Bell test with high enough Bell violation and low enough QBER to be in the regime where the secret key rate is positive. That a *loophole-free* violation of a Bell inequality is indeed a desirable goal was argued, for example, by Gerhardt et al. [18], who showed that it is possible to achieve a violation of a Bell inequality in a system that manifestly lacks entanglement by ignoring either the locality or the detection loophole.

It took until 2015 for the first loophole-free Bell test to be achieved. The first group to report such an experiment was Hensen et al. in Delft [31], who used a novel entanglement-swapping scheme to overcome both loopholes at the same time. The idea of their scheme is depicted in Fig. 7.11 and goes as follows: Alice and Bob each prepare an isolated atom, for example, in a cavity or an ion trap. To

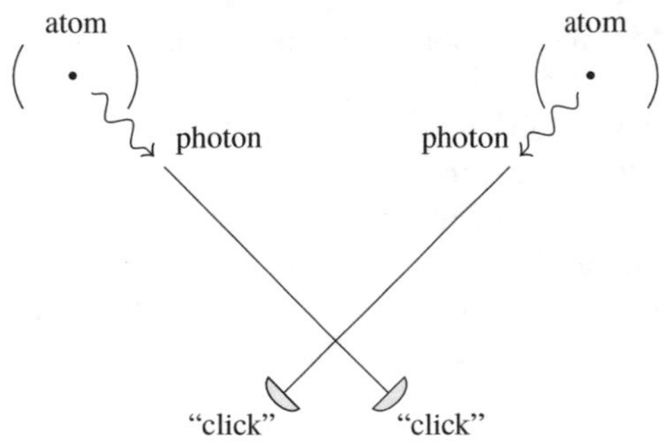

Fig. 7.11 Implementation of a QND via entanglement swapping. Alice and Bob each create atom-photon entanglement in their respective labs. The photons are used to perform an entanglement-swapping procedure such that Alice and Bob share entangled atoms afterwards. These can be used to perform a Bell test

be precise, the Delft group has used electron spins in nitrogen vacancy centres in diamond in laboratories that were set about 1.4 km apart from each other. To get these electron spins to talk to each other, Alice and Bob make them communicate via photons. A stimulated electron sometimes emits a photon, with which it is entangled. In rare cases, the two electron spins even emit a photon at the same time. When this happens, it is possible to send these photons through fibres to perform a procedure called *entanglement swapping*, which entangles the electron spins by measuring the photons. As a result, Alice and Bob now hold entangled electron spins which they can use to perform a Bell test. The crucial point here is that the Bell test is only performed if the entanglement-swapping procedure has been successful (i.e., if both detectors have clicked when measuring the photons). In this way, entanglement swapping is a way to implement a QND, which allows to close both the locality loophole (because the respective labs are far enough apart) and the detection loophole. With this technique the group was able to achieve a CHSH value of 2.38 ± 0.14.

However, the probability that the entanglement-swapping procedure is successful is very small (around 6.4×10^{-9} in the actual experiment, which corresponds to slightly more than one successful procedure per hour), which in turn means that it takes a lot of time to collect enough data for meaningful results. The main reason for the small success probability is the high photon loss due to the long distance between the devices. Together with a QBER of ≈ 0.06, the resulting secret key rate for DIQKD is just at the border of the zero-key region, as depicted in Fig. 7.12.

In the same year, two experiments with entangled photons were reported that also achieved a loophole-free Bell test, one by Giustina et al. [24] and the other by Shalm et al. [68]. However, even though they were able to close both loopholes at the same time, the resulting Bell violation was negligible in both experiments: The first one achieved a CHSH value of $S = 2.000030 \pm 0.000002$, while the second one achieved a CHSH value of $S = 2.00004 \pm 0.00001$. Together with a non-zero QBER, the resulting secure key rate for DIQKD is indeed zero (see Fig. 7.12).

In 2017, another experiment achieving a loophole-free Bell test was reported by Rosenfeld et al. [65] with a CHSH violation of $S = 2.221 \pm 0.033$. An overview of the estimated parameters relevant for DIQKD in experimental Bell tests in recent years is given in Table 7.3. The parameters are plotted in Fig. 7.12 together with the bound for secure key rates in case of collective attacks given in (6.29). This diagram shows that even though loophole-free Bell tests have improved in the past years, they are still not in the region where it is possible to produce a non-zero secret key.

MDI QKD

The first demonstration of a successful implementation of the MDI QKD scheme was achieved in 2013 by Rubenok at al. in Calgary, Canada [66] and by Liu et al. in Hefei, China [42]. The current record regarding the achievable distance with MDI QKD is 404 km with a low-loss fibre, demonstrated in 2016 by Yin et al. [80] with a key rate of 0.00034 bit/s.

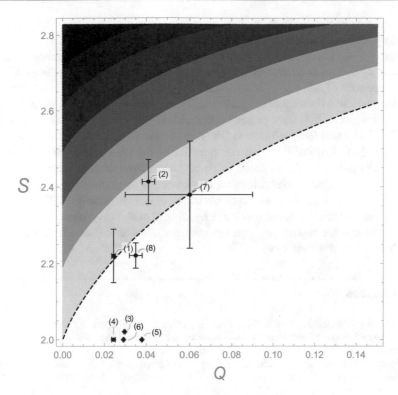

Fig. 7.12 Experimental Bell tests. The contour plot shows the secure key rate r as a function of the Bell violation S and the QBER Q. The location of eight experimental Bell tests is marked (the data can be found in Table 7.3). The dashed line marks the boundary for the zero-key-rate region from (6.29) [1]. Only experiments (5)–(7) close both the detection and the locality loophole

Table 7.3 Summary of experimental Bell tests. The table lists the estimated parameters of interest for a DIQKD protocol, namely the Bell violation S and the QBER Q. Experiments (1) and (2) have used trapped ions, (3)–(6) are all-photonic experiments, (7) used NV centres, and (8) used trapped atoms. All experiments close the detection loophole, but only (5)–(8) additionally close the locality loophole. The data is taken from [53]

Label	Experiment	Year	S	Q
(1)	Matsukevich et al. [51]	2008	2.22 ± 0.07	0.041 ± 0.003
(2)	Pironio et al. [58]	2010	2.414 ± 0.058	0.041 ± 0.003
(3)	Giustina et al. [23]	2013	2.02096 ± 0.00032	0.0297 ± 0.0003
(4)	Christensen et al. [8]	2013	2.00022 ± 0.00003	0.0244 ± 0.0009
(5)	Giustina et al. [24]	2015	2.000030 ± 0.000002	0.0379 ± 0.0002
(6)	Shalm et al. [68]	2015	2.00004 ± 0.00001	0.0292 ± 0.0002
(7)	Hensen et al. [31]	2015	2.38 ± 0.14	0.06 ± 0.03
(8)	Rosenfeld et al. [65]	2017	2.221 ± 0.033	0.035 ± 0.003

Continuous-Variable QKD

An experimental realization of the GG02 protocol we have presented in the previous section is given in [36]. Here, the authors demonstrate secure key generation over a distance of 80 km. More recently, Zhang et al. [84] demonstrated secure CV QKD over a distance of 202.81 km of ultralow-loss optical fibre.

There has also been some progress with regard to cost-efficiency of implementations: Zhang et al. [83] have presented a photonic chip platform for CV QKD using the GG02 protocol that realizes a stable, miniaturized and low-cost implementation of a CV QKD system. Here, all components of the experiment (except the laser source) are put on a silicon photonic chip. The authors then demonstrate a secret key rate of 0.14 kbit/s under collective attacks over a simulated distance of 100 km.

As in DV QKD, there are also considerations of implementing satellite QKD with continuous variables. A feasibility study by Dequal et al. [11] recently showed that it is possible to achieve positive secret key rates for satellite-to-ground communication in a low-earth orbit scenario.

References

1. Acín, A., Brunner, N., Gisin, N., Massar, S., Pironio, S., Scarani, V.: Device-independent security of quantum cryptography against collective attacks. Phys. Rev. Lett. **98**(23) (2007). https://doi.org/10.1103/physrevlett.98.230501
2. Bennett, C.H., Brassard, G.: Experimental quantum cryptography: the dawn of a new era for quantum cryptography: the experimental prototype is working! ACM SIGACT News **20**(4), 78–80 (1989). https://doi.org/10.1145/74074.74087
3. Bennett, C.H., Bessette, F., Brassard, G., Salvail, L., Smolin, J.: Experimental quantum cryptography. J. Cryptol. **5**(1), 3–28 (1992). https://doi.org/10.1007/bf00191318
4. Boaron, A., Boso, G., Rusca, D., Vulliez, C., Autebert, C., Caloz, M., Perrenoud, M., Gras, G., Bussières, F., Li, M.J., Nolan, D., Martin, A., Zbinden, H.: Secure quantum key distribution over 421 km of optical fiber. Phys. Rev. Lett. **121**(19) (2018). https://doi.org/10.1103/physrevlett.121.190502
5. Brassard, G., Lütkenhaus, N., Mor, T., Sanders, B.C.: Limitations on practical quantum cryptography. Phys. Rev. Lett. **85**(6), 1330–1333 (2000). https://doi.org/10.1103/physrevlett.85.1330
6. Braunstein, S.L., Pirandola, S.: Side-channel-free quantum key distribution. Phys. Rev. Lett. **108**(13) (2012). https://doi.org/10.1103/physrevlett.108.130502
7. Braunstein, S.L., van Loock, P.: Quantum information with continuous variables. Rev. Mod. Phys. **77**(2), 513–577 (2005). https://doi.org/10.1103/revmodphys.77.513
8. Christensen, B.G., McCusker, K.T., Altepeter, J.B., Calkins, B., Gerrits, T., Lita, A.E., Miller, A., Shalm, L.K., Zhang, Y., Nam, S.W., Brunner, N., Lim, C.C.W., Gisin, N., Kwiat, P.G.: Detection-Loophole-free test of quantum nonlocality, and applications. Phys. Rev. Lett. **111**(13) (2013). https://doi.org/10.1103/physrevlett.111.130406
9. Curty, M., Xu, F., Cui, W., Lim, C.C.W., Tamaki, K., Lo, H.K.: Finite-key analysis for measurement-device-independent quantum key distribution. Nat. Commun. **5**(1) (2014). https://doi.org/10.1038/ncomms4732
10. Curty, M., Azuma, K., Lo, H.K.: Simple security proof of twin-field type quantum key distribution protocol. NPJ Quantum Inf. **5**(1) (2019). https://doi.org/10.1038/s41534-019-0175-6

11. Dequal, D., Vidarte, L.T., Rodriguez, V.R., Vallone, G., Villoresi, P., Leverrier, A., Diamanti, E.: Feasibility of satellite-to-ground continuous-variable quantum key distribution. NPJ Quantum Inf. **7**(1) (2021). https://doi.org/10.1038/s41534-020-00336-4

12. Diamanti, E., Leverrier, A.: Distributing secret keys with quantum continuous variables: principle, security and implementations. Entropy **17**(12), 6072–6092 (2015). https://doi.org/10.3390/e17096072

13. Diamanti, E., Lo, H.K., Qi, B., Yuan, Z.: Practical challenges in quantum key distribution. NPJ Quantum Inf. **2**(1) (2016). https://doi.org/10.1038/npjqi.2016.25

14. Epping, M., Kampermann, H., Macchiavello, C., Bruß, D.: Multi-partite entanglement can speed up quantum key distribution in networks. New J. Phys. **19**(9), 093012 (2017). https://doi.org/10.1088/1367-2630/aa8487

15. Furrer, F., Franz, T., Berta, M., Leverrier, A., Scholz, V.B., Tomamichel, M., Werner, R.F.: Continuous variable quantum key distribution: finite-key analysis of composable security against coherent attacks. Phys. Rev. Lett. **109**(10) (2012). https://doi.org/10.1103/physrevlett.109.100502

16. García-Patrón, R., Cerf, N.J.: Unconditional optimality of gaussian attacks against continuous-variable quantum key distribution. Phys. Rev. Lett. **97**(19) (2006). https://doi.org/10.1103/physrevlett.97.190503

17. Gerhardt, I., Liu, Q., Lamas-Linares, A., Skaar, J., Kurtsiefer, C., Makarov, V.: Full-field implementation of a perfect eavesdropper on a quantum cryptography system. Nat. Commun. **2**(1) (2011). https://doi.org/10.1038/ncomms1348

18. Gerhardt, I., Liu, Q., Lamas-Linares, A., Skaar, J., Scarani, V., Makarov, V., Kurtsiefer, C.: Experimentally faking the violation of Bell's inequalities. Phys. Rev. Lett. **107**(17) (2011). https://doi.org/10.1103/physrevlett.107.170404

19. Gerry, C., Knight, P.: Introductory Quantum Optics. Cambridge University Press, Cambridge (2010)

20. Ghorai, S., Grangier, P., Diamanti, E., Leverrier, A.: Asymptotic security of continuous-variable quantum key distribution with a discrete modulation. Phys. Rev. X **9**(2) (2019). https://doi.org/10.1103/physrevx.9.021059

21. Gisin, N., Ribordy, G., Tittel, W., Zbinden, H.: Quantum cryptography. Rev. Mod. Phys. **74**(1), 145–195 (2002). https://doi.org/10.1103/revmodphys.74.145

22. Gisin, N., Fasel, S., Kraus, B., Zbinden, H., Ribordy, G.: Trojan-horse attacks on quantum-key-distribution systems. Phys. Rev. A **73**(2) (2006). https://doi.org/10.1103/physreva.73.022320

23. Giustina, M., Mech, A., Ramelow, S., Wittmann, B., Kofler, J., Beyer, J., Lita, A., Calkins, B., Gerrits, T., Nam, S.W., Ursin, R., Zeilinger, A.: Bell violation using entangled photons without the fair-sampling assumption. Nature **497**(7448), 227–230 (2013). https://doi.org/10.1038/nature12012

24. Giustina, M., Versteegh, M.A., Wengerowsky, S., Handsteiner, J., Hochrainer, A., Phelan, K., Steinlechner, F., Kofler, J., Larsson, J.Å., Abellán, C., Amaya, W., Pruneri, V., Mitchell, M.W., Beyer, J., Gerrits, T., Lita, A.E., Shalm, L.K., Nam, S.W., Scheidl, T., Ursin, R., Wittmann, B., Zeilinger, A.: Significant-Loophole-free test of Bell's theorem with entangled photons. Phys. Rev. Lett. **115**(25) (2015). https://doi.org/10.1103/physrevlett.115.250401

25. Gottesmann, D., Lo, H.K., Lütkenhaus, N., Preskill, J.: Security of quantum key distribution with imperfect devices. Quant. Inf. Comput. **4**(5), 325–360 (2004)

26. Grasselli, F.: Quantum Cryptography. Springer International Publishing, Cham (2021). https://doi.org/10.1007/978-3-030-64360-7

27. Grasselli, F., Kampermann, H., Bruß, D.: Finite-key effects in multipartite quantum key distribution protocols. New J. Phys. **20**(11), 113014 (2018). https://doi.org/10.1088/1367-2630/aaec34

28. Grosshans, F., Grangier, P.: Continuous variable quantum cryptography using coherent states. Phys. Rev. Lett. **88**(5) (2002). https://doi.org/10.1103/physrevlett.88.057902

29. Grosshans, F., Assche, G.V., Wenger, J., Brouri, R., Cerf, N.J., Grangier, P.: Quantum key distribution using gaussian-modulated coherent states. Nature **421**(6920), 238–241 (2003). https://doi.org/10.1038/nature01289

30. Grosshans, F., Cerf, N.J., Wenger, J., Tualle-Brouri, R., Grangier, P.: Virtual entanglement and reconciliation protocols for quantum cryptography with continuous variables. Quantum Info. Comput. **3**(7), 535–552 (2003)
31. Hensen, B., Bernien, H., Dréau, A.E., Reiserer, A., Kalb, N., Blok, M.S., Ruitenberg, J., Vermeulen, R.F.L., Schouten, R.N., Abellán, C., Amaya, W., Pruneri, V., Mitchell, M.W., Markham, M., Twitchen, D.J., Elkouss, D., Wehner, S., Taminiau, T.H., Hanson, R.: Loophole-free Bell inequality violation using electron spins separated by 1.3 kilometres. Nature **526**(7575), 682–686 (2015). https://doi.org/10.1038/nature15759
32. Hillery, M.: Quantum cryptography with squeezed states. Phys. Rev. A **61**(2) (2000). https://doi.org/10.1103/physreva.61.022309
33. Hong, C.K., Ou, Z.Y., Mandel, L.: Measurement of subpicosecond time intervals between two photons by interference. Phys. Rev. Lett. **59**(18), 2044–2046 (1987). https://doi.org/10.1103/physrevlett.59.2044
34. Inamori, H.: Security of practical time-reversed EPR quantum key distribution. Algorithmica **34**(4), 340–365 (2002). https://doi.org/10.1007/s00453-002-0983-4
35. Joshi, S.K., Aktas, D., Wengerowsky, S., Lončarić, M., Neumann, S.P., Liu, B., Scheidl, T., Lorenzo, G.C., Samec, Ž., Kling, L., Qiu, A., Razavi, M., Stipčević, M., Rarity, J.G., Ursin, R.: A trusted node–free eight-user metropolitan quantum communication network. Sci. Adv. **6**(36), eaba0959 (2020). https://doi.org/10.1126/sciadv.aba0959
36. Jouguet, P., Kunz-Jacques, S., Leverrier, A., Grangier, P., Diamanti, E.: Experimental demonstration of long-distance continuous-variable quantum key distribution. Nat. Photonics **7**(5), 378–381 (2013). https://doi.org/10.1038/nphoton.2013.63
37. Leverrier, A., Grangier, P.: Unconditional security proof of long-distance continuous-variable quantum key distribution with discrete modulation. Phys. Rev. Lett. **102**(18) (2009). https://doi.org/10.1103/physrevlett.102.180504
38. Leverrier, A., García-Patrón, R., Renner, R., Cerf, N.J.: Security of continuous-variable quantum key distribution against general attacks. Phys. Rev. Lett. **110**(3) (2013). https://doi.org/10.1103/physrevlett.110.030502
39. Liao, S.K., Cai, W.Q., Liu, W.Y., Zhang, L., Li, Y., Ren, J.G., Yin, J., Shen, Q., Cao, Y., Li, Z.P., Li, F.Z., Chen, X.W., Sun, L.H., Jia, J.J., Wu, J.C., Jiang, X.J., Wang, J.F., Huang, Y.M., Wang, Q., Zhou, Y.L., Deng, L., Xi, T., Ma, L., Hu, T., Zhang, Q., Chen, Y.A., Liu, N.L., Wang, X.B., Zhu, Z.C., Lu, C.Y., Shu, R., Peng, C.Z., Wang, J.Y., Pan, J.W.: Satellite-to-ground quantum key distribution. Nature **549**(7670), 43–47 (2017). https://doi.org/10.1038/nature23655
40. Liao, S.K., Cai, W.Q., Handsteiner, J., Liu, B., Yin, J., Zhang, L., Rauch, D., Fink, M., Ren, J.G., Liu, W.Y., Li, Y., Shen, Q., Cao, Y., Li, F.Z., Wang, J.F., Huang, Y.M., Deng, L., Xi, T., Ma, L., Hu, T., Li, L., Liu, N.L., Koidl, F., Wang, P., Chen, Y.A., Wang, X.B., Steindorfer, M., Kirchner, G., Lu, C.Y., Shu, R., Ursin, R., Scheidl, T., Peng, C.Z., Wang, J.Y., Zeilinger, A., Pan, J.W.: Satellite-relayed intercontinental quantum network. Phys. Rev. Lett. **120**(3) (2018). https://doi.org/10.1103/physrevlett.120.030501
41. Lin, J., Upadhyaya, T., Lütkenhaus, N.: Asymptotic security analysis of discrete-modulated continuous-variable quantum key distribution. Phys. Rev. X **9**(4) (2019). https://doi.org/10.1103/physrevx.9.041064
42. Liu, Y., Chen, T.Y., Wang, L.J., Liang, H., Shentu, G.L., Wang, J., Cui, K., Yin, H.L., Liu, N.L., Li, L., Ma, X., Pelc, J.S., Fejer, M.M., Peng, C.Z., Zhang, Q., Pan, J.W.: Experimental measurement-device-independent quantum key distribution. Phys. Rev. Lett. **111**(13) (2013). https://doi.org/10.1103/physrevlett.111.130502
43. Lo, H.K., Curty, M., Qi, B.: Measurement-device-independent quantum key distribution. Phys. Rev. Lett. **108**(13) (2012). https://doi.org/10.1103/physrevlett.108.130503
44. Lo, H.K., Curty, M., Tamaki, K.: Secure quantum key distribution. Nat. Photonics **8**(8), 595–604 (2014). https://doi.org/10.1038/nphoton.2014.149
45. Lucamarini, M., Yuan, Z.L., Dynes, J.F., Shields, A.J.: Overcoming the rate–distance limit of quantum key distribution without quantum repeaters. Nature **557**(7705), 400–403 (2018). https://doi.org/10.1038/s41586-018-0066-6

46. Lydersen, L., Wiechers, C., Wittmann, C., Elser, D., Skaar, J., Makarov, V.: Hacking commercial quantum cryptography systems by tailored bright illumination. Nat. Photonics **4**(10), 686–689 (2010). https://doi.org/10.1038/nphoton.2010.214

47. Ma, X., Zeng, P., Zhou, H.: Phase-matching quantum key distribution. Phys. Rev. X **8**(3) (2018). https://doi.org/10.1103/physrevx.8.031043

48. Makarov, V.: Controlling passively quenched single photon detectors by bright light. New J. Phys. **11**(6), 065003 (2009). https://doi.org/10.1088/1367-2630/11/6/065003

49. Makarov, V., Skaar, J.: Faked states attack using detector efficiency mismatch on SARG04, phase-time, DPSK, and Ekert protocols. Quantum Inf. Comput. **8**(6&7), 0622–0635 (2008)

50. Makarov, V., Anisimov, A., Skaar, J.: Effects of detector efficiency mismatch on security of quantum cryptosystems. Phys. Rev. A **74**(2) (2006). https://doi.org/10.1103/physreva.74.022313

51. Matsukevich, D.N., Maunz, P., Moehring, D.L., Olmschenk, S., Monroe, C.: Bell inequality violation with two remote atomic qubits. Phys. Rev. Lett. **100**(15) (2008). https://doi.org/10.1103/physrevlett.100.150404

52. Matsuura, T., Maeda, K., Sasaki, T., Koashi, M.: Finite-size security of continuous-variable quantum key distribution with digital signal processing. Nat. Commun. **12**(1) (2021). https://doi.org/10.1038/s41467-020-19916-1

53. Murta, G., van Dam, S.B., Ribeiro, J., Hanson, R., Wehner, S.: Towards a realization of device-independent quantum key distribution. Quantum Sci. Technol. **4**(3), 035011 (2019). https://doi.org/10.1088/2058-9565/ab2819

54. Murta, G., Grasselli, F., Kampermann, H., Bruß, D.: Quantum conference key agreement: a review. Adv. Quantum Technol. **3**(11), 2000025 (2020). https://doi.org/10.1002/qute.202000025

55. Navascués, M., Grosshans, F., Acín, A.: Optimality of Gaussian attacks in continuous-variable quantum cryptography. Phys. Rev. Lett. **97**(19) (2006). https://doi.org/10.1103/physrevlett.97.190502

56. Pirandola, S., Laurenza, R., Ottaviani, C., Banchi, L.: Fundamental limits of repeaterless quantum communications. Nat. Commun. **8**(1) (2017). https://doi.org/10.1038/ncomms15043

57. Pirandola, S., Andersen, U.L., Banchi, L., Berta, M., Bunandar, D., Colbeck, R., Englund, D., Gehring, T., Lupo, C., Ottaviani, C., Pereira, J., Razavi, M., Shaari, J.S., Tomamichel, M., Usenko, V.C., Vallone, G., Villoresi, P., Wallden, P.: Advances in quantum cryptography. Adv. Opt. Photonics **12**(4), 1012–1236 (2019). https://doi.org/10.1364/AOP.361502

58. Pironio, S., Acín, A., Massar, S., de la Giroday, A.B., Matsukevich, D.N., Maunz, P., Olmschenk, S., Hayes, D., Luo, L., Manning, T.A., Monroe, C.: Random numbers certified by Bell's theorem. Nature **464**(7291), 1021–1024 (2010). https://doi.org/10.1038/nature09008

59. Qi, B., Fung, C.H.F., Lo, H.K., Ma, X.: Time-shift attack in practical quantum cryptosystems. Quantum Inf. Comput. **7**(1&2), 073–083 (2007)

60. Ralph, T.C.: Continuous variable quantum cryptography. Phys. Rev. A **61**(1) (1999). https://doi.org/10.1103/physreva.61.010303

61. Ralph, T.C.: Security of continuous-variable quantum cryptography. Phys. Rev. A **62**(6) (2000). https://doi.org/10.1103/physreva.62.062306

62. Rarity, J.G., Tapster, P.R., Gorman, P.M., Knight, P.: Ground to satellite secure key exchange using quantum cryptography. New J. Phys. **4**, 82–82 (2002). https://doi.org/10.1088/1367-2630/4/1/382

63. Reid, M.D.: Quantum cryptography with a predetermined key, using continuous-variable Einstein-Podolsky-Rosen correlations. Phys. Rev. A **62**(6) (2000). https://doi.org/10.1103/physreva.62.062308

64. Renner, R., Cirac, J.I.: de Finetti representation theorem for infinite-dimensional quantum systems and applications to quantum cryptography. Phys. Rev. Lett. **102**(11) (2009). https://doi.org/10.1103/physrevlett.102.110504

65. Rosenfeld, W., Burchardt, D., Garthoff, R., Redeker, K., Ortegel, N., Rau, M., Weinfurter, H.: Event-Ready Bell test using entangled atoms simultaneously closing detection and locality Loopholes. Phys. Rev. Lett. **119**(1) (2017). https://doi.org/10.1103/physrevlett.119.010402

66. Rubenok, A., Slater, J.A., Chan, P., Lucio-Martinez, I., Tittel, W.: Real-world two-photon interference and proof-of-principle quantum key distribution immune to detector attacks. Phys. Rev. Lett. **111**(13) (2013). https://doi.org/10.1103/physrevlett.111.130501
67. Sangouard, N., Simon, C., de Riedmatten, H., Gisin, N.: Quantum repeaters based on atomic ensembles and linear optics. Rev. Mod. Phys. **83**(1), 33–80 (2011). https://doi.org/10.1103/revmodphys.83.33
68. Shalm, L.K., Meyer-Scott, E., Christensen, B.G., Bierhorst, P., Wayne, M.A., Stevens, M.J., Gerrits, T., Glancy, S., Hamel, D.R., Allman, M.S., Coakley, K.J., Dyer, S.D., Hodge, C., Lita, A.E., Verma, V.B., Lambrocco, C., Tortorici, E., Migdall, A.L., Zhang, Y., Kumor, D.R., Farr, W.H., Marsili, F., Shaw, M.D., Stern, J.A., Abellán, C., Amaya, W., Pruneri, V., Jennewein, T., Mitchell, M.W., Kwiat, P.G., Bienfang, J.C., Mirin, R.P., Knill, E., Nam, S.W.: Strong Loophole-free test of local realism. Phys. Rev. Lett. **115**(25) (2015). https://doi.org/10.1103/physrevlett.115.250402
69. Sych, D., Leuchs, G.: Coherent state quantum key distribution with multi letter phase-shift keying. New J. Phys. **12**(5), 053019 (2010). https://doi.org/10.1088/1367-2630/12/5/053019
70. Takenaka, H., Carrasco-Casado, A., Fujiwara, M., Kitamura, M., Sasaki, M., Toyoshima, M.: Satellite-to-ground quantum-limited communication using a 50-kg-class microsatellite. Nat. Photonics **11**(8), 502–508 (2017). https://doi.org/10.1038/nphoton.2017.107
71. Takeoka, M., Guha, S., Wilde, M.M.: Fundamental rate-loss tradeoff for optical quantum key distribution. Nat. Commun. **5**(1) (2014). https://doi.org/10.1038/ncomms6235
72. Tamaki, K., Lo, H.K., Wang, W., Lucamarini, M.: Information theoretic security of quantum key distribution overcoming the repeaterless secret key capacity bound (2018). arXiv. Preprint. https://arxiv.org/abs/1805.05511
73. Vakhitov, A., Makarov, V., Hjelme, D.R.: Large pulse attack as a method of conventional optical eavesdropping in quantum cryptography. J. Mod. Opt. **48**(13), 2023–2038 (2001). https://doi.org/10.1080/09500340108240904
74. Weedbrook, C., Lance, A.M., Bowen, W.P., Symul, T., Ralph, T.C., Lam, P.K.: Quantum cryptography without switching. Phys. Rev. Lett. **93**(17) (2004). https://doi.org/10.1103/physrevlett.93.170504
75. Weedbrook, C., Pirandola, S., García-Patrón, R., Cerf, N.J., Ralph, T.C., Shapiro, J.H., Lloyd, S.: Gaussian quantum information. Rev. Mod. Phys. **84**(2), 621–669 (2012). https://doi.org/10.1103/revmodphys.84.621
76. Weier, H., Krauss, H., Rau, M., Fürst, M., Nauerth, S., Weinfurter, H.: Quantum eavesdropping without interception: an attack exploiting the dead time of single-photon detectors. New J. Phys. **13**(7), 073024 (2011). https://doi.org/10.1088/1367-2630/13/7/073024
77. Winzer, P.J.: Scaling optical fiber networks: challenges and solutions. Opt. Photon. News **26**(3), 28 (2015). https://doi.org/10.1364/opn.26.3.000028
78. Wolf, M.M., Giedke, G., Cirac, J.I.: Extremality of Gaussian quantum states. Phys. Rev. Lett. **96**(8) (2006). https://doi.org/10.1103/physrevlett.96.080502
79. Xu, F., Ma, X., Zhang, Q., Lo, H.K., Pan, J.W.: Secure quantum key distribution with realistic devices. Rev. Mod. Phys. **92**(2) (2020). https://doi.org/10.1103/revmodphys.92.025002
80. Yin, H.L., Chen, T.Y., Yu, Z.W., Liu, H., You, L.X., Zhou, Y.H., Chen, S.J., Mao, Y., Huang, M.Q., Zhang, W.J., Chen, H., Li, M.J., Nolan, D., Zhou, F., Jiang, X., Wang, Z., Zhang, Q., Wang, X.B., Pan, J.W.: Measurement-device-independent quantum key distribution over a 404 km optical fiber. Phys. Rev. Lett. **117**(19) (2016). https://doi.org/10.1103/physrevlett.117.190501
81. Yin, J., Li, Y.H., Liao, S.K., Yang, M., Cao, Y., Zhang, L., Ren, J.G., Cai, W.Q., Liu, W.Y., Li, S.L., Shu, R., Huang, Y.M., Deng, L., Li, L., Zhang, Q., Liu, N.L., Chen, Y.A., Lu, C.Y., Wang, X.B., Xu, F., Wang, J.Y., Peng, C.Z., Ekert, A.K., Pan, J.W.: Entanglement-based secure quantum cryptography over 1,120 kilometres. Nature **582**(7813), 501–505 (2020). https://doi.org/10.1038/s41586-020-2401-y

82. Yuan, Z., Murakami, A., Kujiraoka, M., Lucamarini, M., Tanizawa, Y., Sato, H., Shields, A.J., Plews, A., Takahashi, R., Doi, K., Tam, W., Sharpe, A.W., Dixon, A.R., Lavelle, E., Dynes, J.F.: 10-Mb/s quantum key distribution. J. Light. Technol. **36**(16), 3427–3433 (2018). https://doi.org/10.1109/jlt.2018.2843136

83. Zhang, G., Haw, J.Y., Cai, H., Xu, F., Assad, S.M., Fitzsimons, J.F., Zhou, X., Zhang, Y., Yu, S., Wu, J., Ser, W., Kwek, L.C., Liu, A.Q.: An integrated silicon photonic chip platform for continuous-variable quantum key distribution. Nat. Photonics **13**(12), 839–842 (2019). https://doi.org/10.1038/s41566-019-0504-5

84. Zhang, Y., Chen, Z., Pirandola, S., Wang, X., Zhou, C., Chu, B., Zhao, Y., Xu, B., Yu, S., Guo, H.: Long-distance continuous-variable quantum key distribution over 202.81 km of fiber. Phys. Rev. Lett. **125**(1) (2020). https://doi.org/10.1103/physrevlett.125.010502

85. Zhao, Y., Fung, C.H.F., Qi, B., Chen, C., Lo, H.K.: Quantum hacking: experimental demonstration of time-shift attack against practical quantum-key-distribution systems. Phys. Rev. A **78**(4) (2008). https://doi.org/10.1103/physreva.78.042333

Basic Probability Theory

A

In order to follow many of the arguments in these notes, especially when talking about entropies, it is necessary to have some basic knowledge of probability theory. Therefore, we review here the most important tools of probability theory that are used.

One of the basic notions of probability theory that also frequently appears throughout these notes is that of a discrete *random variable*. A random variable X can take one of several values, the so-called *realizations* x, given by the *alphabet* X. The probability that a certain realization $x \in X$ occurs is given by the probability distribution $p_X(x)$. We usually use upper case letters to denote the random variable, lower case letters to denote realizations thereof, and calligraphic letters to denote the alphabet.

Suppose we have two random variables X and Y, which may depend on each other. We can then define the *joint probability distribution* $p_{X,Y}(x, y)$ of X and Y that tells you the probability that $Y = y$ and $X = x$. This notion (and the following definition) can be expanded to n random variables, but we restrict ourselves to the case of pairs X, Y here to keep the notation simple.

Given the joint probability distribution of the pair X, Y, we can derive the *marginal distribution* $P_X(x)$ by

$$p_X(x) = \sum_{y \in \mathcal{Y}} p_{X,Y}(x, y) \qquad \forall x \in X \tag{A.1}$$

and analogously for $P_Y(y)$. The two random variables X and Y are said to be *independent* if

$$p_{X,Y}(x, y) = p_X(x) p_Y(y). \tag{A.2}$$

© The Author(s), under exclusive license to Springer Nature Switzerland AG 2021
R. Wolf, *Quantum Key Distribution*, Lecture Notes in Physics 988,
https://doi.org/10.1007/978-3-030-73991-1

Furthermore, we can define the *conditional probability* that Y takes the value $y \in \mathcal{Y}$, given that X takes the value $x \in \mathcal{X}$:

$$p_{Y|X}(y|x) = \frac{p_{X,Y}(x, y)}{p_X(x)}. \tag{A.3}$$

To avoid complications, we use the convention that $p_{X,Y}(x, y) = 0$ if $p_X(x) = 0$. If X and Y are independent, $p_{Y|X}(x|y) = p_Y(y)$ for all $y \in \mathcal{Y}$. Using the definition of the conditional probability, (A.1) can be rewritten as

$$p_X(x) = \sum_{y \in \mathcal{Y}} p_{X|Y}(x|y) p_Y(y) \quad \forall x \in \mathcal{X}. \tag{A.4}$$

In this form it is also called the *law of total probability*. Another important rule that relates different conditional probabilities is *Bayes' rule*:

$$p_{X|Y}(x|y) = p_{Y|X}(y|x) \frac{p_X(x)}{p_Y(y)}. \tag{A.5}$$

This rule can be proved as follows: Note that (A.3) can be rewritten as

$$p_{X,Y}(x, y) = p_{Y|X}(y|x) p_X(x). \tag{A.6}$$

It follows that

$$p_{X|Y}(x|y) = \frac{p_{X,Y}(x, y)}{p_Y(y)} = p_{Y|X}(y|x) \frac{p_X(x)}{p_Y(y)}. \tag{A.7}$$

Calderbank–Shor–Steane Codes

B

Calderbank–Shor–Steane (CSS) codes are a large class of quantum error correction codes that exploit ideas from classical linear error correction codes. In entanglement-based QKD protocols, they can be used to correct errors that occur during the distribution of entangled states.

B.1 Classical Linear Codes

Before we can understand CSS codes, we need to make a short detour into the theory of classical linear codes. A *linear code C* that encodes k bits into an n bit code space (with $n > k$) is a set of 2^k codewords, where each codeword is a binary vector of length n. We call such a code an $[n, k]$ code. It is specified by a $n \times k$ *generator matrix G* with elements in $\{0, 1\}$. G maps messages to their equivalent in the code space, for instance, a k bit message x (which is represented by a column vector) is encoded as $y = Gx$. Note that all arithmetic operations (especially multiplications and additions) are done modulo 2.

As a simple example, consider the $[3, 1]$ repetition code that encodes 1 bit messages into three copies of them: 0 is mapped to $(0, 0, 0)^T$ and 1 is mapped to $(1, 1, 1)^T$. Hence, the generator matrix G is

$$G = \begin{pmatrix} 1 \\ 1 \\ 1 \end{pmatrix}. \tag{B.1}$$

To connect this definition of classical codes to error correction, we have to introduce a different formulation of linear codes, the *parity check* matrices. In this formulation, an $[n, k]$ code is defined as all vectors x of length n with entries from $\{0, 1\}$ such that

$$Hx = 0, \tag{B.2}$$

© The Author(s), under exclusive license to Springer Nature Switzerland AG 2021
R. Wolf, *Quantum Key Distribution*, Lecture Notes in Physics 988,
https://doi.org/10.1007/978-3-030-73991-1

where H is an $(n - k) \times n$ matrix with entries in $\{0, 1\}$ called the parity check matrix. To construct the parity matrix H from a generator matrix G, one has to pick out $n - k$ linearly independent vectors orthogonal to the columns of G. The corresponding parity check matrix for the $[3, 1]$ repetition code with G given in (B.1) is then

$$H = \begin{pmatrix} 1 & 1 & 0 \\ 0 & 1 & 1 \end{pmatrix}. \tag{B.3}$$

In the language of parity check matrices, it is quite easy to see how error detection and correction work. Suppose we have a message x that we encode as $y = Gx$. If an error e occurs, the codeword y is transformed into the corrupted codeword $y' = y + e$. Because $Hy = 0$ for all codewords y, it follows that $Hy' = Hy + He = He$. This is called the *error syndrome*. If the syndrome is 0, we know that no error has occurred. Otherwise, it contains information about the error because of the way the parity check matrix H was constructed.

In the example of the $[3, 1]$ repetition code, every codeword has a length of 3 bits. Therefore, errors can occur at three different positions. Denote by e_i an error in the ith bit, i.e., a vector with a 1 at position i. Then for all codewords y, we have that $Hy' = He_i$; hence, the three different syndromes are

$$He_1 = \begin{pmatrix} 1 \\ 0 \end{pmatrix}, \qquad He_2 = \begin{pmatrix} 0 \\ 1 \end{pmatrix}, \qquad He_3 = \begin{pmatrix} 1 \\ 1 \end{pmatrix}. \tag{B.4}$$

This makes it possible to read off the position of the error from the syndromes. Note that this procedure is only successful if we know that an error has occurred for at most one bit. Hence, the $[3, 1]$ repetition code can correct one error.

More general linear error correction codes can be obtained using the concepts of *Hamming distance*. The Hamming distance $d(x, y)$ between two binary vectors x and y is defined as the number of positions in which the two bit strings differ. For example, $d((1, 1, 0, 0)^T, (1, 0, 0, 1)^T) = 2$, because the vectors differ in the 2nd and 4th positions. Error correction now works as follows: suppose we have a codeword $y = Gx$ that is corrupted such that the resulting vector is $y' = y + e$. If the probability that an error occurs is less than $\frac{1}{2}$, the most likely codeword to have been encoded is the one that minimizes the Hamming distance to y', i.e., $d(y, y')$, since this is the one with the least amount of bit flips.

How many errors can such a code correct? This can also be analysed in terms of the Hamming distance: We define the distance of a code C to be the minimum Hamming distance between any of its codewords:

$$d(C) = \min_{x, y \in C, \, x \neq y} d(x, y). \tag{B.5}$$

We use the notation $d = d(C)$ and call C an $[n, k, d]$ code. With a little bit of thinking one can see that a code with distance $2t + 1$ for some integer t can be used to correct up to t errors, simply by decoding the corrupted message y' as the *unique* codeword y that satisfies $d(y, y') \leq t$. If more than t errors occur, this codeword is no longer unique and therefore, errors cannot be reliably detected and corrected.

The last concept we need from classical linear codes is *duality*. Suppose we have a linear $[n, k]$ code C with generator matrix G and parity check matrix H. We can then construct another code, the dual code $C^{\perp}$ of C, which consists of all codewords that are orthogonal to each codeword in C. Hence, the generator matrix of the dual code is H^T and its parity check matrix is G^T.

B.2 Quantum Error Correction

In the quantum case, the situation is a bit more complicated. Where in the classical case only one type of error is possible (namely the bit flip error), a qubit can undergo three different types of errors: a bit flip, which changes $|0\rangle$ to $|1\rangle$ and $|1\rangle$ to $|0\rangle$, a phase error, which maps $|1\rangle$ to $-|1\rangle$ but leaves $|0\rangle$ unchanged, and a combination of the two, which maps $|0\rangle \rightarrow -|1\rangle$ and $|1\rangle \rightarrow |0\rangle$.

The Calderbank–Shor–Steane (CSS) code is now defined as follows: Suppose we have two classical linear error correction codes, an $[n, k_1]$ code C_1 and an $[n, k_2]$ code C_2 such that $C_2 \subset C_1$ and C_1 and $C_2^{\perp}$ both correct up to t errors. Using these two classical codes we can define a quantum error correction code, the *CSS code of C_1 over C_2*, denoted $\text{CSS}(C_1, C_2)$. It is an $[n, k_1 - k_2]$ quantum code that is capable of correcting errors on up to t qubits. The construction works as follows: for any codeword $x \in C_1$, we define the quantum state

$$|x + C_2\rangle = \frac{1}{\sqrt{|C_2|}} \sum_{y \in C_2} |x + y\rangle, \tag{B.6}$$

where $+$ is the bitwise addition modulo 2 and $|C_2|$ denotes the cardinality of C_2 (which is 2^{k_2}, since this is the number of codewords of C_2).

We have used coset notation here for a reason. If you are not familiar with the concept of a coset, we briefly recap some facts here for a group G and a subgroup $H \subset G$, then for any $g \in G$ the *left coset* of H in G determined by g is defined as

$$g + H = \{g + h : h \in H\}. \tag{B.7}$$

We denote G/H the set of all cosets of H in G. Cosets have some convenient properties: Coming back to the notation for CSS codes, suppose that x' is an element of C_1 such that $x - x' \in C_2$. Then it follows that $|x + C_2\rangle = |x' + C_2\rangle$, which implies that the state $|x + C_2\rangle$ only depends on the coset of C_1/C_2 in which x is contained. In this sense, (B.6) is an equally weighted superposition of all the words in the coset represented by x.

Another consequence of the coset formalism is that if x and x' belong to different cosets of C_2 in C_1, then there are no codewords $y, y' \in C_2$ such that $x + y = x' + y'$;

hence, $|x+C_2\rangle$ and $|x'+C_2\rangle$ are orthonormal states. The quantum code $\mathrm{CSS}(C_1, C_2)$ is defined to be the vector space spanned by $\{|x + C_1\rangle\}_{x \in C_1}$. Since the number of cosets of C_2 in C_1 is $|C_1|/|C_2|$, the dimension of this vector space is $|C_1|/|C_2| = 2^{k_1 - k_2}$, and therefore $\mathrm{CSS}(C_1, C_2)$ is an $[n, k_1 - k_2]$ quantum code.

It is now possible to exploit the classical error-correcting properties of the codes C_1 and $C_2^{\perp}$ to detect and correct quantum errors. The crucial point here is that bit flip errors and phase flip errors are corrected independent of each other. Bit flip errors are described by a vector e_{bit} of length n that has 1s at those positions where a bit flip has occurred, and 0s otherwise. If the original state is denoted $|x\rangle$, bit flip errors transform this state to

$$|x\rangle \rightarrow |x + e_{\mathrm{bit}}\rangle. \tag{B.8}$$

Phase errors are described by a second vector e_{phase} of length n with 1s at those positions where a phase error has occurred. In this case, the phase errors transform the state $|x\rangle$ as

$$|x\rangle \rightarrow (-1)^{x \cdot e_{\mathrm{phase}}} |x\rangle. \tag{B.9}$$

A crucial observation here is that when we apply the Hadamard transformation (see (2.28)), the phase error takes the same form as the bit flip error,[1] i.e., a state $|x'\rangle$ in the Hadamard basis is transformed by phase errors as

$$|x'\rangle \rightarrow |x' + e_{\mathrm{phase}}\rangle. \tag{B.10}$$

In summary, if $|x + C_2\rangle$ as defined in (B.6) is the original state, then the corrupted state is described as

$$\frac{1}{\sqrt{2^{k_2}}} \sum_{y \in C_2} (-1)^{(x+y) \cdot e_{\mathrm{phase}}} |x + y + e_{\mathrm{bit}}\rangle. \tag{B.11}$$

To detect bit flip errors, we need to compute the error syndrome for the code C_1. For this purpose it is convenient to introduce an ancilla state that consists of a sufficient number of qubits to store the syndrome and that initially is in the all zero state $|0\rangle$. To compute the syndrome, we apply the parity check matrix H_1 of the code C_1 and store the result in the ancilla state:

$$|x+y+e_{\mathrm{bit}}\rangle|0\rangle \rightarrow |x+y+e_{\mathrm{bit}}\rangle|H_1(x+y+e_{\mathrm{bit}})\rangle = |x+y+e_{\mathrm{bit}}\rangle|H_1 e_{\mathrm{bit}}\rangle. \tag{B.12}$$

Hence, to detect the error one simply measures the ancilla state, discards it, and applies NOT gates (i.e., gates that take $|0\rangle \rightarrow |1\rangle$ and $|1\rangle \rightarrow |0\rangle$) to those qubits

---

[1]One can easily verify this statement by carrying out this computation for the two basis states for the Hadamard basis, namely $|+\rangle$ and $|-\rangle$ defined in (2.24).

where a bit flip has occurred. This removes all the bit flip errors and the resulting state is

$$\frac{1}{\sqrt{2^{k_2}}} \sum_{y \in C_2} (-1)^{(x+y)\cdot e_{\text{phase}}} |x + y\rangle. \tag{B.13}$$

The remaining part is to detect and correct phase errors. We can do this by applying a Hadamard transformation to each qubit, which transforms the state to

$$\frac{1}{\sqrt{2^{n+k_2}}} \sum_{z} \sum_{y \in C_2} (-1)^{(x+y)\cdot(e_{\text{phase}}+z)} |z\rangle, \tag{B.14}$$

where the sum is over all possible n bit values for z. We can rewrite this state by setting $z' = z + e_{\text{phase}}$, which yields

$$\frac{1}{\sqrt{2^{n+k_2}}} \sum_{z'} \sum_{y \in C_2} (-1)^{(x+y)\cdot z'} |z' + e_{\text{phase}}\rangle. \tag{B.15}$$

One can show that if $z' \in C^\perp$, then $\sum_{y \in C_2} (-1)^{y \cdot z'} = |C_2|$, while if $z' \notin C^\perp$, then $\sum_{y \in C_2} (-1)^{y \cdot z'} = 0$, which allows us to further rewrite the state:

$$\frac{1}{\sqrt{2^{n-k_2}}} \sum_{z' \in C_2^\perp} (-1)^{x \cdot z'} |z' + e_{\text{phase}}\rangle, \tag{B.16}$$

which has exactly the form of a bit flip error described by the vector e_{phase}. We can therefore simply repeat the procedure we did before, but now with the parity check matrix of the code $C_2^\perp$. Here, it becomes clear why we need $C_2^\perp$ to be able to correct t errors and not C_2 itself. This allows us to correct all the errors and we receive the state

$$\frac{1}{\sqrt{2^{n-k_2}}} \sum_{z' \in C_2^\perp} (-1)^{x \cdot z'} |z'\rangle. \tag{B.17}$$

The last step is to apply the Hadamard transformation again to each qubit since it is its own inverse. The resulting state is

$$\frac{1}{\sqrt{2^{k_2}}} \sum_{y \in C_2} |x + y\rangle, \tag{B.18}$$

which is exactly the originally encoded state.

Index

© The Author(s), under exclusive license to Springer Nature Switzerland AG 2021 227
R. Wolf, *Quantum Key Distribution*, Lecture Notes in Physics 988,
https://doi.org/10.1007/978-3-030-73991-1

Printed in the United States
by Baker & Taylor Publisher Services